Foundations of Engineering Mechanics

Series Editors: V.I. Babitsky, J. Wittenburg

Liudmila Ya. Banakh · Mark L. Kempner

Vibrations of Mechanical Systems with Regular Structure

With 109 Figures

Springer

Dr. Liudmila Ya. Banakh
Mechanical Engineering Research Institute
Mal. Kharitonjevsky Str. 4
101990 Moscow
Russia
banl@inbox.ru

Prof. Mark L. Kempner
Ezra Str. 24/5
76201 Rehovot
Israel
kempner@netvision.net.il

Series Editors
V.I. Babitsky
Department of Mechanical Engineering
Loughborough University
LE11 3TU Loughborough
Leicestershire
Great Britain

J. Wittenburg
Universitat Karlsruhe (TH)
Institute für Technishe
Mechanik
Kaiserstr. 12
76128 Karlsruhe
Germany

ISSN 1612-1384 e-ISSN 1860-6237
ISBN 978-3-642-03125-0 e-ISBN 978-3-642-03126-7
DOI 10.1007/978-3-642-03126-7
Springer Heidelberg Dordrecht London New York

Library of Congress Control Number: 2010928905

Cover design: deblik, Berlin

Printed on acid-free paper

Springer is part of Springer Science+Business Media (www.springer.com)

Preface

In this book, regular structures are defined as periodic structures consisting of repeated elements (translational symmetry) as well as structures with a geometric symmetry. Regular structures have for a long time been attracting the attention of scientists by the extraordinary beauty of their forms. They have been studied in many areas of science: chemistry, physics, biology, etc. Systems with geometric symmetry are used widely in many areas of engineering. The various kinds of bases under machines, cyclically repeated forms of stators, reduction gears, rotors with blades mounted on them, etc. represent regular structures.

The study of real-life engineering structures faces considerable difficulties because they comprise a great number of working mechanisms that, in turn, consist of many different elastic subsystems and elements. The computational models of such systems represent a hierarchical structure and contain hundreds and thousands of parameters. The main problems in the analysis of such systems are the dimension reduction of model and revealing the dominant parameters that determine its dynamics and form its energy nucleus.

The two most widely used approaches to the simulation of such systems are as follows:

1. Methods using lumped parameters models, i.e., a discretization of the original system and its representation as a system with lumped parameters [including finite-element method (FEM)].
2. The use of idealized elements with distributed parameters and known analytical solutions for both the local elements and the subsystems.

Each of these approaches has its own specific characteristics and methods of study, which are described in Chap. 2. On this basis, the book contains two parts.

The first part is devoted to the study of vibrations in linear systems with lumped (concentrated) parameters. In this part, dispersion equations (for systems with a periodic structure) as well as the theory of groups representation (for systems with geometric symmetry) are used. However, the specifics of the mechanical systems required a certain generalization of these approaches: Generalized projective operators of symmetry were introduced. This made it possible to take into account the symmetry hierarchy in system, the small asymmetry arising as a result of the technological errors. The interactions of natural modes at vibroisolation of a body mounted on a symmetric frame and also the dynamic properties of a planetary reduction gear were investigated. The aforementioned approaches permit one to carry out the significant numerical simplification, using the characteristics of only one cell, and by this way to obtain analytical results.

The second part is devoted to systems with distributed parameters. In this part, the methods of dynamic compliances (stiffness) proposed by Professor M. L. Kempner are developed. In these approaches as form-building cells the idealized elements with distributed parameters are used. Methods for calculating such systems have been worked out. The basic matrices describing these systems, such as the transition matrix and the mixed matrix, as well as equations in finite differences, have been determined. With the help of these approaches, it is possible to obtain results in analytical form even for such complex elements of mechanical engineering structures as turbine disks with a blades package, multimass rotors, and cylindrical shells with circular and longitudinal ribbing. Both symmetric elements and reflection symmetry elements were treated.

Self-similar structures (fractals), in which every consecutive element is formed from the preceding one by a rotation and/or similarity transformation (scaling), can also be added to the regular structures. Examples of such systems in mechanical engineering are crankshafts, rods and beams whose parameters change in the same ratio from section to section, and the rotors of compressors where the disks are connected by means of conic shells.

The dynamic properties of such mechanical structures have not been studied thoroughly enough until now. A number of their dynamic properties are provided in this book. In particular, it is shown that some classes of self-similar structures represent a mechanical band filter. The natural frequencies of such systems as elements of flying machines e.g. rotors with self-similar disks and rotors of the drum type have been determined.

The book contains many numerical examples for real-life engineering structures, in particular aviation engines.

Part I was written by L.Y. Banakh, Part II by M.L. Kempner, and Sect. 3.4 by P.S. Akhmetkhanov.

This book is based on the original research the authors. The authors would like to thank Prof. V. Babitsky for his attention to this work.

Moscow, Russia Liudmila Ya. Banakh
Rehovot, Israel Mark L. Kempner

Contents

Chapter 1
Introduction

A broad class of systems in nature and engineering is represented by regular structures consisting of repeated elements or having a geometrical symmetry. A significant amount of research in various fields of science is devoted to the study of such systems.

The ideal nature of the buildup of regular structures predetermines their specific dynamic properties. So, for example, systems with repeated elements (with a periodic structure) possess filtering properties and have frequency condensation points.

The particular attractiveness of such structures consists in the possibility of using only a single form-generating cell in their study. Therefore, in many cases, it is possible to obtain an analytical solution even for rather complex systems with a hierarchical structure. A particular prominence is given in this book to obtaining analytical results.

However, mechanical regular structures represent a special class and have a number of specific features that differentiate them from structures created by nature.

Above all, as a rule, mechanical regular structures are subsystems of large structures and in turn consist of rather complex subsystems. As a result, their computational models have a large number of degrees of freedom and parameters. The choice a computational model of a system is ambiguous and does not represent a formalized procedure. The most important objective consists in revealing in this model the basic dominant elements that determine the main dynamic properties of the system. The choice of a model determines the methods of its study.

In the first part of the book, the vibrations of regular systems with lumped parameters and finite-element models are discussed. We represent also algorithm for derivation of FE-model for beam system in analytical form. The second part is devoted to the vibrations in systems with distributed parameters.

Various methods for studying symmetrical systems have been known for a long time in the theory of oscillations. The ideas for resolving the forces and displacements for elementary types of symmetry are developed in the theory of engineering structures [44, 100, 102], etc. In [29], a method based on the spectral features of cyclical matrices is used. In [37] the problem of the vibrations of a bar system is treated with the methods of structural mechanics and group representation theory. Here to every kind of displacements in a system corresponds its own subsystem

L.Y. Banakh, M.L. Kempner, *Vibrations of Mechanical Systems with Regular Structure*, Foundations of Engineering Mechanics, DOI 10.1007/978-3-642-03126-7_1,

with special boundary conditions. In [115], the method of G-invariant subspaces is used in relation to the static of structures. Group representation theory received further development in [26], where the elementary cell is used. In [33], the concept of "equivalent points" for solids included in a mathematical model was introduced. This permits one to increase considerably the symmetry level of a system. It has also been shown in a number of studies that the presence of nonconservative forces narrows the symmetry level.

The models with distributed parameters for the systems with cyclic symmetry are discussed in [29, 45, 46, 102]. Some general approaches to the study of regular systems can be found in [80]. An interesting aspect appeared to be the transition from equations in finite differences to a continuous medium, i.e., the continualization problem [69, 70].

The most universal and elegant method in the study of systems with geometric symmetry is the theory of groups representation. The methods of this theory make it possible to obtain information about the behavior of a physical system without solving its equations, analyzing only its symmetry type. This is of particular importance when these equations are completely unknown. G. Weil wrote the following: "Each time when it is necessary to deal with some object, try to determine its automorphic group. This approach makes it possible to penetrate deeply into its internal properties." And in fact, each type of symmetry is related to a certain conservation law and, consequently, to a corresponding invariant of systems.

In the first part, we use theory of representation of discrete groups [61, 71]. This theory differs from the theory of continuous group Lee [84, 113], which uses the symmetry of partial differential equations. The mathematical apparatus of theory of groups representation was first proposed by Evariste Galois (1831). This theory is used successfully in physics and chemistry for studying the properties of crystals and multinuclear molecules. Its mathematical apparatus can be applied in the study of systems with any type of symmetry. It is difficult to overestimate the advantages of this approach. It makes it possible to determine easily and exhaustively the peculiarities of a system, including its dynamic features.

However, the specifics of mechanical systems required a certain generalization of these approaches. That is why we introduced generalized projective operators of symmetry and generalized modes. These concepts make it possible to take into account the hierarchy of subsystems and their multidimensionality. They also allow one to compress information in a large system with a hierarchical structure. This approach is very promising due to its algorithmic nature and the possibility of obtaining analytical results with the help of sufficiently simple matrix transformations.

The problems related to violations of the regularity as well as the stability are also closely related to the problems of regular structures. These problems can be caused by a small asymmetry of the structure, which usually emerges in mechanical structures. Then the systems become quasi-symmetrical. The applying generalized projective operators and decomposition approaches, taking into account weak dynamic interactions, also make it possible to solve this problem. The generalized modes which are invariants for given symmetry type are found for frame systems, for vibroisolation of a body on a symmetric frame, and for the system of a planetary reduction gear. It enable the vibrations interactions between elements and subsystems to analyze.

The next class of the discussed systems is represented by systems with repeated elements (systems with translational symmetry). A large contribution to the development of systems with such a structure in mechanical engineering was made by works on the statics and dynamics of structures [6, 22, 23, 35, 36, 55, 80, 89, 110]. Thereby, the method of forces (displacements), which is traditional for the theory of structures, as well as a mixed method is used.

The wave approach [25] is traditionally used in studying the oscillations of periodic systems. This approach consists in obtaining a dispersion equation that links the wavelength and frequencies of the free vibration. The use of dispersion equations provides a lot of information about a complex system and – which is very important – in a condensed form. These equations reflect the general feature of the oscillating structure under any boundary conditions. This makes it possible to obtain a solution for free and for forced oscillations of an assembled system (or its separate parts) by using equations for only one cell. Therefore, we solve equations whose order is equal to the number of the degrees of freedom for one form-shaping element.

This approach is widely used in quantum mechanics for studying the properties of molecules [89, 91, 97]. In acoustics, it is used for studying extended structures in the high-frequency range when the wavelength is considerably less than the dimensions of the structure under investigation [78]. In these cases, it is possible to treat the system as endless.

It was used [6, 78] for studying structures in the mechanical engineering in the high-frequency range, in particular for systems with angular junctures, discrete inclusions, and ribbings. The dispersion equations were obtained with the help of the dynamic compliance for the construction elements. The reception of dispersion equations was used in [48] for solving problems in aerodynamics.

We use the dispersion equations for finite-element models. It has enabled to obtain the analytical results for wide class of mechanical systems such as beams systems, frames, laminar systems with sparsely positioned laminar ribbing and also for FE-models of symmetrical systems.

Regular structures with translational symmetry represent nonetheless a relatively limited class of structures and do not describe many natural and technical systems. However, this can be expanded considerably by complementing them with a class of self-similar structures (fractals). The fractal is a structure consisting of parts that are similar to the structure as a whole [32, 75, 76]. This is one of the most important principles of form building in nature. In the field of mechanical engineering, they include, for example, a beam with a stepwise change in its cross section, crankshafts (screw symmetry), a drum-type rotor with a conic shell, etc.

The specific characteristics of such systems in the process of their vibrations have not been studied sufficiently. We analyze some dynamical peculiarities in this book. In particular, it has been shown that a self-similar beam is a mechanical banded filter.

Its pass band is inversely related to the scale coefficient. When the scale coefficient is large, the system can be considered a one-frequency filter.

As it was above mentioned the dynamic analysis of mechanical structures including many subsystems is effective only when the system is treated as an assembled. When studying large systems of any nature, the most powerful and natural approaches are decomposition ones. These approaches allow one to analyze the subsystems separately and then to take into account the effect of the interaction between these subsystems. This is very efficient both for the theoretical and numerical analysis. The methods of decomposition of dynamic systems are developing at present in the following directions:

a) Variants of the component synthesis of normal modes;
b) Topological methods using graph theory based on the works of G. Kron "Diakoptics" [67];

The topological approaches use graph theory, the tensor calculus, and also electromechanical analogies. These methods have received a widespread prevalence and are used in various fields of science: informatics, mechanical engineering, and economics. Initially they were used in electric circuit theory; however, afterwards they became widely used in automatic control theory, computerization, etc. G. Kron [67] uses two information sources about a system: its equations and its graphs. For the assembling subsystems he uses the subgraphs, introduction of additional nodes, elimination of existing ones, etc. In the afterword to Russian translation of the book, A. V. Baranov proved that these procedures are equivalent to simple matrix transformations: elimination of coordinates, introduction of additional nodes, permutation of rows and columns, etc.

In [92], topological methods are used for mechanical systems. The use of these methods here requires further generalizations due to the multidimensionality of the graph branches, the presence of reciprocal elastic connections between elements. The subsystems assembling were performed with help of compliance matrices.

The methods of network theory are also widely used in the analysis of large mechanical systems, including rotor systems [34, 101]. In [108], graphs and oriented graphs are used in the equivalent transformation of vibration models.

But it must be pointed out that in these methods the process of decomposition into subsystems is arbitrary and does not take into account certain specific features of the system. However, the arbitrary decomposition of the mathematical model is not that efficient. Actually, the solution for an assembled system differs considerably from that for individual systems. That is why we use an approach that takes into consideration the level of dynamic interactions between subsystems. For these purpose the coefficients determining the value of dynamic interactions in multibody mechanical systems are given [11].

It was found that weak interactions emerge, in particular, at wide diapason of elastic-inertial parameters, at a small asymmetry in system, with strong frequency disparity between the subsystems [9, 10]. And these factors practically always are available in the big systems. Therefore it is the main property of hierarchical mechanical systems.

The vibrations in weakly interacted subsystems take place practically independently of each other. In this case the decomposition of the physical system and its mathematical model coincide. The use of various decomposition ways permits to reveal the dominant coordinates in given frequency range.

The systems with distributed parameters we consider in the second part of the book.

The methods of dynamic compliances (stiffness) proposed by Kempner [53] are very efficient in the study of rather complex real-life structures. Here these methods are essentially generalized. In this case, the dynamic model is presented in the form of a set of elements with distributed parameters. For each element of this kind, the dynamic compliances (stiffness) are determined as a harmonic function of the excitation frequency. Since these quantities depend on the excitation frequency, they are called dynamic compliances (stiffness) by analogy with static compliances (stiffness). When analyzing the vibrations of two linked subsystems, the compatibility condition at the point of linking must be fulfilled. This condition means that sum of dynamic compliances for subsystems is equal to zero. The frequency that makes the sum equal to zero will be the natural frequency of the assembled system.

This approach can be easily generalized for systems with damping by introducing a complex frequency describing the damping [30].

However, the possibilities of the proposed methods are considerably wider than the merely static analogy. The fact that the dynamic stiffness (compliance) of an element can be expressed through its static stiffness and the spectrum of its natural frequencies is very important for practical use. This opens up a wide possibility for its experimental determination and for the subsequent incorporation of the given element into the main system [53, 54]. The preliminary linearization of the element is also possible. This is particularly important for real-life engineering structures when the analytic studies are extremely complicated.

The proposed methods solve the problem of determining the dependence between the dynamic characteristics of an assembled system and its formative elements. In this sense, it can be considered that this position is a variant of the above-mentioned Diakoptics [67] – the analysis of complex systems in parts.

These methods turned out to be efficient enough in the study of the vibrations of systems with a regular structure. It makes effective use of equations in finite differences [64, 82], transition and mixed matrices.

There is also the investigation of regular systems with reflection-symmetric and antisymmetric elements as well as those with self-similar elements. The proposed approaches allow one to find a solution in analytical form. In addition, the vibrations of cylindrical shells with round stiffness ribbings at arbitrary boundary conditions, as well as vibrations of cylindrical shells with longitudinal stiffness ribbings, including the asymmetric profile, are studied.

In book we use mainly the matrix methods. The matrix methods become particularly valuable in the study of linear oscillations in large systems. They are convenient not only because of the easy numerical calculations. The structure of the matrix reflects the physical structure of system and clearly describes the interactions between the subsystems and the elements. The studies [5, 27, 31, 34, 86] have made a large contribution to the problem of applying the matrix methods in the theory of oscillations.

From a mathematical point of view, the methods that have been discussed can be treated as a transformation of coordinates or as a space transformation. Such a transformation is possible not only for regular systems but also in the case of self-similar structures.

On the basis of the proposed approaches, studies of the vibrations in real-life structures have been performed, such as: rotors with a regular and self-similar structure; vibroisolation systems on symmetrical frames; elements of flying machines, such as shrouded blades; reduction gear systems.

Chapter 2
Mechanical Vibratory Systems with Hierarchical Structure. Simulation and Calculation Methods

2.1 Introduction

Real-life machines, such as gas-turbine and liquid propellant rocket engines, machine tools, etc., comprise a large number of working mechanisms. In turn, these mechanisms consist of various elastic subsystems and components. The study of the dynamic processes in such systems can lead to exhaustive results only when the system is treated as a whole entity and the interactions between the all substructures and elements are taken into consideration. The computational models of such systems with a hierarchical structure contain hundreds and thousands of parameters. The cardinal problem in this respect is above all the simulation of a large system and then the revealing of a relatively small number of variables that determine its dynamics and constitute its energy "nucleus."

The methods of vibration analysis of mechanical systems with a hierarchical structure have been in active development since the 1940s and 1950s. The dynamic behavior of large systems is very versatile. The possibility for describing these or other properties depends essentially on the accepted computational model and the calculation methods. At present, there are many different approaches to the simulation of a real structure and its vibrational analysis. They can be consolidated into several large groups:

1. Methods using the models with lumped parameters. This is a discretization of the original system and its representation as a system with lumped parameters (including the finite-element method (FEM)).
2. Using as a model idealized elements with distributed parameters and a known analytical solution.
3. Continualization, i.e., creating a computational model in the form of a continuous medium. This means a certain smoothing of the sufficiently densely positioned discrete inclusions and elements [49, 70]. In this way, a continuous model of a system with distributed parameters is created; it is described by partial differential equations. This method is particularly effective for the high-frequency range. In this case, a model with distributed parameters, equivalent to the initial system at the energy level, can be deriving [85]. There are asymptotic estimates for the

L.Y. Banakh, M.L. Kempner, *Vibrations of Mechanical Systems with Regular Structure*, Foundations of Engineering Mechanics, DOI 10.1007/978-3-642-03126-7_2,

vibrational spectrum [109, v. 1]. At present, the so-called methods of statistical energy analysis have also received wide prevalence.

The methods of idealization of the first two groups, which permit one to study vibrations of the mechanical systems in a sufficiently wide frequency range, are used in this book. The vibrations of linear systems are also discussed.

2.2 Models of Mechanical Systems with Lumped Parameters

In a real-life structure, there are usually at the same time elements that can be treated as concentrated masses or solids and also as elements with distributed parameters. The sufficiently rigid massive inclusions could be assigned to the first group, while the beams, plates, shells, etc. to the second one.

There are different ways of discretization. The representation of systems in the form of concentrated masses or solids connected by massless elastic elements is one of the earliest methods. The examples in Fig. 2.1 show simplified models as systems with lumped parameters for a number of structures.

The computational model describing the plane bending vibrations of the rotor of a turbo-driven pump aggregate can be simulated in the form shown in Fig. 2.1a. This is a system with several concentrated masses positioned in the inertia centers of the disks and connected to each other by massless connections. The long rocket structure can be simulated as a free system with dot masses connected by elastic beams (Fig. 2.1b). A gas turbine engine with the casing taken into consideration is presented in Fig. 2.1c. The casing here is simulated in the form of rigid masses connected by elastic rods. The elastic elements comprise the connections between the rotor and the casing and between the casing and the engine support points. They all

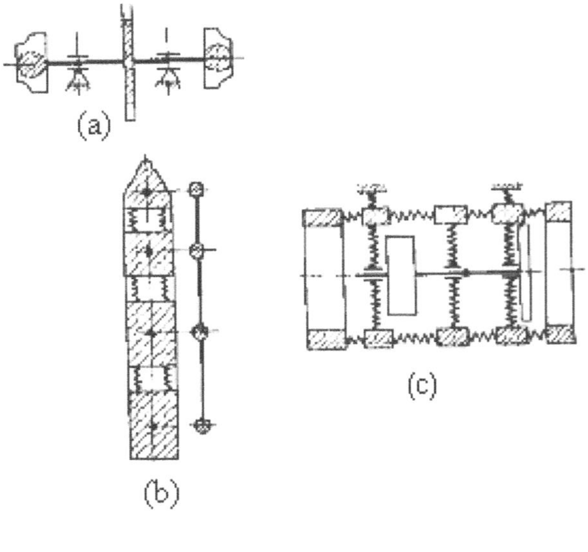

Fig. 2.1a–c Discrete models of various mechanical systems: (**a**) rotor of a turbo-driven pump aggregate, (**b**) components of a rocket, (**c**) gas turbine engine with casing

are simulated as weightless elastic elements whose stiffness is equal to the stiffness of the corresponding elements in the basic system.

Such a method of simulating the structure describes sufficiently well the low-frequency vibrations. However, the problem of the accuracy of the model depending on the number of selected masses remains open. It is necessary to keep in mind that refinement of the model performed with help of very small elements can lead to an incorrect result if the wavelength of the element is commensurate with the cross-section size [20].

Discrete models are described by ordinary differential equations. In this case, the elements of the structure, such as massive disks or heavy sections of shafts or beams, are treated as rigid bodies with the same masses and inertia moments. Very often they are replaced by dot masses concentrated in the inertia center of the corresponding rigid body. But in this case the angular displacements of the concentrated masses cannot be taken into account.

Discrete models are described by ordinary differential equations in two ways: by equations of the following kind:

$$m_s\ddot{x}_s + \sum_{j=1}^{n} k_{sj}x_j = Q_s(t), \tag{2.1}$$

or by the following equations:

$$x_s + \sum_{j=1}^{n} m_j\delta_{sj}\ddot{x}_j = \sum_{j=1}^{n} \delta_{sj}Q_j(t) \quad (s, j = 1\ldots n). \tag{2.2}$$

In Eq. (2.1), k_{sj} are the static stiffness coefficients, m_s is the mass concentrated in node s, and $Q_s(t)$ is the external force applied to node s. In Eq. (2.2), δ_{sj} are the static compliance coefficients.

2.2.1 Coefficients of Static Stiffness and Compliance

Static stiffness coefficients. Let us now determine the physical meaning of the static stiffness coefficients. In order to do that, let us first consider the following problem of statics. Let us set in (2.1) the inertia terms $m_i = 0\,(i = 1\ldots n)$ and the force $Q = \text{const}$.

$$\sum_{j=1}^{n} k_{sj}x_j = Q_s. \tag{2.3}$$

Now it is clear from (2.3) that the static stiffness coefficients in any node, for example in s are defined as

$$k_{js} = \frac{Q_j}{x_s} \text{ at } x_j \equiv 0\,(j = 1\ldots n,\ j \neq s). \tag{2.4}$$

In such a way, this coefficient is numerically equal to the force emerging in the node j at a unit displacement of node s. At the same time, all other $n - 1$ nodes $(s \neq j)$ are rigidly fixed.

At $s = j$, the coefficient of stiffness k_{ss} is named as principal (or main), at $s \neq j$ it is named as collateral (mixed).

According to Betty's reciprocity theorem: $k_{sj} = k_{js}$.

For the free vibrations we have in matrix form

$$(-\mathbf{M}\Lambda + \mathbf{K})\mathbf{X}_0 = 0,$$

$$\underset{(nxn)}{\mathbf{K}} = \begin{bmatrix} k_{11} & k_{21} & \ldots & k_{n1} \\ k_{21} & k_{22} & \ldots & k_{2n} \\ \ldots & \ldots & \ldots & \ldots \\ k_{n1} & k_{n2} & \ldots & k_{nn} \end{bmatrix} \quad \text{is a stiffness matrix.}$$

The inertia matrix $\mathbf{M}$ is often set as a diagonal matrix, i.e.,

$$\mathbf{M} = diag[m_i] = \begin{bmatrix} m_1 & & \\ & \ddots & \\ & & m_n \end{bmatrix}.$$

$$\Lambda = diag[\omega_s^2], \ \omega_s \text{ are natural frequencies}$$

The spectrum of natural frequencies ω_s as solution of eigenvalue problem is defined

$$\det(-\mathbf{M}\Lambda + \mathbf{K}) = 0\,,$$

For harmonic excitation with frequency ω we have

$$Q_s = Q_{s0} \sin \omega t, \quad x_s = x_{s0} \sin \omega t \ (s = 1 \ldots n).$$

Then Eq. (2.1) in matrix form has the view

$$(-\mathbf{M}\omega^2 + \mathbf{K})\mathbf{X}_0 = \mathbf{Q}_0(t). \tag{2.5}$$

$\mathbf{X}_0$ denotes a column vector of the vibrations amplitudes for coordinates x_s, $\mathbf{Q}_0$ is a column vector of amplitudes for the external forces. Equation (2.5) has the meaning of a compatibility condition for the forces in the nodes.

After taking into account the viscous linear damping, new terms $b_i\dot{x}_i$, depending on the velocity, will appear in Eq. (2.5). In this case, by setting $x_r = A_r\, e^{i\omega t}$, we find that the matrix

$$\mathbf{D}\,(\omega) = (-\mathbf{M}\omega^2 + i\mathbf{B}\omega + \mathbf{K})$$

becomes a complex one.

Static compliance coefficients. These are defined as a ratio of the coordinate displacement x_j in node j caused by force Q_s applied to node s:

$$\delta_{js} = \frac{x_j}{Q_s} \quad \text{at } Q_j \equiv 0\,(j \neq s).$$

At the same time in all other nodes the forces are equal to zero.

The matrix of static compliance $\boldsymbol{\delta} = [\delta_{sj}]$. Motion equation (2.2) in matrix form:

$$\mathbf{X} + \boldsymbol{\delta}\mathbf{M}\ddot{\mathbf{X}} = \boldsymbol{\delta}\mathbf{Q}(t).$$

This equation has the meaning of a compatibility condition for displacements in the nodes. It is obvious that

$$\mathbf{K} = \boldsymbol{\delta}^{-1}.$$

But it is necessary to point out that the elements of matrices $\mathbf{K}$ and $\boldsymbol{\delta}$ are not inverse to each other, i.e.,

$$k_{sj} \neq \frac{1}{\delta_{sj}},$$

as they are calculated under different conditions of fixation of the nodes. Their convertibility is possible only in the case of a one-dimensional chain system.

The electrical impedance z_{sj} (conductivity y_{sj}) is determined analogically; in this case, the velocity $\dot{x}_s$ is used instead of the displacement x_i. In the term "admittance," which is usually used in experimental studies, the acceleration $\ddot{x}_s$ is used. These coefficients are convenient for an electromechanical system of analogies.

2.2.2 Static Stiffness Coefficients for a Beam

Many books on the theory of vibrations [20, 106] contain various examples for determining the coefficients of static stiffness and compliance. Nevertheless, we will provide a simple method for determining the static stiffness coefficients for a beam. We will need these relationships later on in determining the stiffness matrix for beam finite element in analytical form.

These coefficients for a beam element can be derived easily from the known equations of theoretical mechanics. Figure 2.2a, b shows a plane beam element [21].

The generalized coordinates are the displacement y and the angle of rotation θ, i.e., the tangent to the bending deflection curve. The bending moment is M_{bend}.

The static equation for a beam is

$$EIy'' = M_{\text{bend}}(y). \tag{2.6}$$

Then we determine all coefficients of stiffness in accordance with Eq. (2.4). For this we set sequentially the unitary displacement on each of the generalized

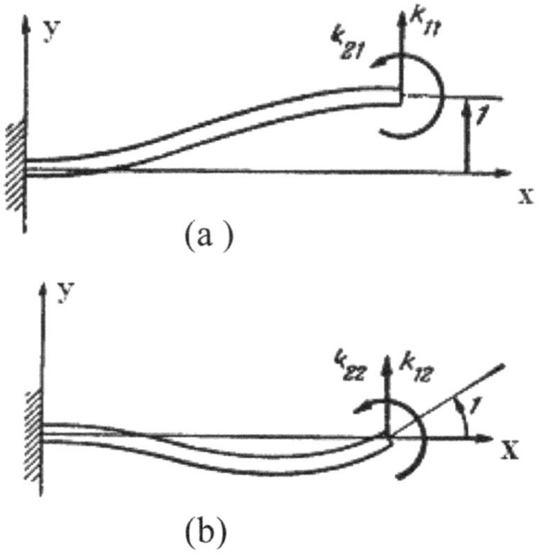

Fig. 2.2a,b Determining the static stiffness coefficients for a plane beam element in the case of transversal (**a**) and angular (**b**) displacements

coordinates and determining the respective reactions in the node. And in addition to that we will set the following boundary conditions:

$$\text{At}\, x = l:\; y = 1;\; \theta_y = 0;\; \text{at}\, x = 0:\; y = 0,\; \theta_y = 0\,.$$

Then the stiffness coefficients k_{11}, k_{21} are equal, respectively, to the force and the moment that have to be applied in order to maintain the beam in the position shown in Fig. 2.2a. Thus it is obvious that

$$M_{bend}(y) = k_{11}(l - y) + k_{21}\,.$$

By integrating the beam Eq. (2.6) at the set boundary conditions, we obtain

$$k_{11} = \frac{12EI}{l^3}, \quad k_{21} = -\frac{6EI}{l^2}\,. \tag{2.6a}$$

Analogically, we determine k_{12}, k_{22}. From Fig. 2.2b

$$k_{22} = \frac{4EI}{l}, \; k_{21} = k_{12}. \tag{2.6b}$$

After the coefficients of stiffness have been determined, the static stiffness matrix for the right-hand end of the beam can be formed:

$$\mathbf{K} = \begin{bmatrix} k_{11} & k_{12} \\ k_{21} & k_{22} \end{bmatrix}.$$

2.3 Reduction of Models with Lumped Parameters

As has already been pointed out, when discretization methods (or FEM) are used, computational models with a great number of degrees of freedom are obtained. That is why the various methods for dimension reduction as well as methods for decomposition of the dynamic model, acquire great importance.

The limiting of the number of generalized coordinates is possible, for example, in the following cases:

- If there are additional connections between the elements;
- If the system contains elements with small inertia terms.

On this basis, the following approaches exist.

Condensation of the stiffness matrix in the case of additional connections. In this case, the condensation of the stiffness matrix can be achieved by eliminating the coordinates with a rigid fixation because there are no displacements along these coordinates [15]. However, the fixable coordinate may not coincide with the selected coordinate axes and can represent a linear combination of the generalized coordinates $x^* = \sum_i \alpha_i x_i$. Then, at the beginning it is necessary to perform a preliminary coordinates transformation in order to delimit coordinate x^* in the initial matrix

$$x^* = \begin{bmatrix} 1 & & & & \\ & 1 & & & \\ \alpha_1 & \alpha_2 & \alpha_3 & \cdots & \alpha_n \\ \cdots & \cdots & \cdots & \cdots & \cdots \\ & & & & 1 \end{bmatrix} x = \mathbf{N} x.$$

After this transformation, the matrix of the system will appear as follows:

$$\mathbf{D^*} = \mathbf{N^{-1} D N}.$$

Now one can eliminate the corresponding row and column from the transformed matrix $\mathbf{D^*}$ and obtain in this way a system of the order $n - 1$.

As is known from the Rayleigh theorem [105] for a system with additional connections, its $n - 1$ eigenfrequencies ω'_i are located between the n eigenfrequencies of the initial system, i.e., $\omega_1 < \omega'_1 < \omega_2 < \omega'_2 < \cdots < \omega_n$. This fact can be used to evaluate the eigenfrequencies through analysis of simpler subsystems formed from the initial system by fixing of some nodes.

As regards the *compliance matrix* for systems with additional connections, their condensation can be achieved by eliminating the rows and columns that correspond to the free nodes because the generalized forces along these coordinates are equal to zero.

Condensation of the stiffness matrix at the existence of the inertia-free elements. In this case, the condensation of the stiffness matrix can be performed by known elimination of the inertia-free elements (Gaussian elimination) [39]. Let us consider Eq. (2.5) at $\mathbf{Q}=0$. Let us assume that the inertia elements corresponding to coordinates $\mathbf{X}_2$ are equal to zero. After regrouping the rows and the columns, we write down the matrix in the form

$$\begin{bmatrix} \mathbf{M}_1 & \\ & \mathbf{0} \end{bmatrix} \begin{bmatrix} \ddot{\mathbf{X}}_1 \\ \ddot{\mathbf{X}}_2 \end{bmatrix} + \begin{bmatrix} \mathbf{K}_{11} & \mathbf{K}_{12} \\ \mathbf{K}_{21} & \mathbf{K}_{22} \end{bmatrix} \begin{bmatrix} \mathbf{X}_1 \\ \mathbf{X}_2 \end{bmatrix} = 0. \quad (2.7)$$

From Eq. (2.7)

$$\mathbf{X}_2 = \mathbf{K}_{21}\mathbf{K}_{22}^{-1}\mathbf{X}_1 \quad (2.8)$$

the equation for the inertia-free coordinates $\mathbf{X}_2$ is converted into an equation of statics, and they are determined from algebraic equation (2.8). From Eqs. (2.7) and (2.8) we obtain a matrix equation for a reduced system with n–l generalized coordinates (l is the dimension of vector $\mathbf{X}_2$):

$$\mathbf{M}_1\ddot{\mathbf{X}}_1 + (\mathbf{K}_1 - \mathbf{K}_3\mathbf{K}_4^{-1}\mathbf{K}_2)\mathbf{X}_1 = \mathbf{0}.$$

Dynamic condensation. If the inertia matrix (2.7) contains nonzero off-diagonal terms, i.e., $\mathbf{M}_{12}, \mathbf{M}_{21}, \mathbf{M}_{22} \neq 0$,

$$\mathbf{M} = \begin{bmatrix} \mathbf{M}_{11} & \mathbf{M}_{12} \\ \mathbf{M}_{21} & \mathbf{M}_{22} \end{bmatrix},$$

then after elimination of the coordinates, the inertia matrix is transformed in the following way:

$$\mathbf{M}^* = \mathbf{M}_1 - \mathbf{K}_{12}\mathbf{K}_{22}^{-1}\mathbf{M}_{22}\mathbf{K}_{22}^{-1}\mathbf{K}_{21} - \mathbf{K}_{21}\mathbf{K}_{22}^{-1}\mathbf{M}_1 - \mathbf{K}_{12}\mathbf{K}_{22}^{-1}\mathbf{M}_{21}. \quad (2.9)$$

***Rayleigh–Ritz method* [17].** In this method, the eigenvector $\mathbf{X}_i$ is searched for in the view of a linear combination of several basic vectors $\mathbf{\Phi}_j$ that satisfy the boundary conditions:

$$\mathbf{X}_i^* = \Sigma_{\alpha_j}\mathbf{\Phi}_j \quad (j = 1 \ldots n, i = 1 \ldots q),\ q < n. \quad (2.10)$$

In this way, $\mathbf{X}^*_j$ depends now on $q < n$ parameters α_j. The success of this method depends primarily on the adequate selection of the basic vectors. Some bending deflections caused by certain loads, e.g., from the action of the gravity force, are usually selected as basic vectors. Then the function $\mathbf{\Phi}_j$ can be found from the equations of statics:

$$\mathbf{K\Phi} = \mathbf{P},$$

$\mathbf{\Phi} = [\Phi_{1\ldots}\ \Phi_n]$ is an orthogonal matrix of the order of n x q, which is formed by q basic functions.

Further on, by treating (2.10) as a transformation of coordinates and setting $\mathbf{K}^* = \mathbf{\Phi}^{\mathbf{T}}\mathbf{K}\mathbf{\Phi} = \mathbf{\Phi}^{\mathbf{T}}\mathbf{P}, \mathbf{M}^* = \mathbf{\Phi}^{\mathbf{T}}\mathbf{M}\mathbf{\Phi},$ we solve the eigenvalue problem for a matrix of the order of q: $(\mathbf{K}^* - \mathbf{\Lambda}^*\mathbf{M}^*) = \mathbf{0}$. As a result, q approximate eigenvalues of $\mathbf{\Lambda}^*$ and the eigenvectors $\mathbf{X}^* = \mathbf{\Phi}\mathbf{X}$ are determined. It follows from the Rayleigh theorem that the approximated eigenvalues are greater than the real ones, i.e., $\Lambda < \mathbf{\Lambda}^*$.

The method of static condensation can be considered as a variant of the Ritz method with basic vectors $\mathbf{\Phi}_j$.

In [7], an approach using the normal modes of Bulgakov in a complex form is elaborated, which does not require any constraints on the type of the damping characteristics and also allows one to take into account the gyroscopic terms.

The method of synthesis of the vibration modes **consists in the following.** First, the vibration modes are determined individually for every subsystem. Then they are synthesized in order to obtain the vibration modes for the assembled system whole. Different types of fixations for the subsystems are used in the process: with free boundaries, with fixed ones, with mixed boundary conditions [19, 62, 66, 72, 96, 99]. In this way, several types of displacement that are compatible with the connections can be taken into account. The different modifications of these approaches use an iteration procedure, approximated or imprecise vibration modes. For example, in [28], the internal and external coordinates for the subsystems are differentiated. Low-level vibration modes are calculated for the internal coordinates with the external ones fixed:

$$(\mathbf{K}_{ss} - \Lambda\mathbf{M}_{ss})\tilde{\mathbf{\Phi}} = 0.$$

The obtained solution is superimposed on the external coordinates:

$$\mathbf{x}_s = -\mathbf{K}_{ss}^{-1}\mathbf{K}_{sm}\mathbf{x}_m + \tilde{\mathbf{\Phi}}\mathbf{y}.$$

These methods of reduction are efficient for systems with a not very large dimension (up to 100 degrees of freedom). In the case of systems with larger dimension, other approaches are used, such as, for example, elimination of the modal coordinates [103], decomposition methods (see 2.7, 2.8).

2.4 Coefficients of Dynamic Stiffness and Compliance

The further generalization of the presented approaches for simulating and calculating mechanical oscillating systems consists in the introduction of coefficients of *dynamic stiffness and compliance*. These coefficients are convenient for the analysis of harmonic vibrations. To this end, the system is divided into elementary subsystems (cells). The coefficients of dynamic stiffness and compliance are determined for these cells. As we will see below, they can be easily determined if the resonance and antiresonance frequencies of the selected cells are known [53]. That is why elements with known analytical solutions (for example, such as the beams, plates, shells, etc.) are selected as cells.

2.4.1 Coefficients of Dynamic Stiffness

Let us go back to Eq. (2.2). In the case of harmonic vibrations, when $x_s = x_{s0} \sin \omega t$, these equations have the form

$$\sum_{j=1}^{n} (-\omega^2 m_s + k_{sj}) x_{s0} = Q_{s0} \quad (s = 1 \ldots n)\,, \tag{2.11}$$

where x_{s0} is the vibration amplitude and Q_{s0} is the amplitude of the exciting force.

The coefficients at coordinates

$$-\omega^2 m_s + k_{sj} = r_{sj}(\omega). \tag{2.12}$$

$r_{sj}(\omega)$ are called coefficients of *dynamic stiffness (rigidity)*. With the help of these coefficients, Eq. (2.11) can be written in a view that is analogous to (2.3)

$$\sum_{j=1}^{n} r_{sj}(\omega) x_{j0} = Q_{s0}(t),$$

or in a matrix form

$$\mathbf{R}(\omega)\,\mathbf{X}_0 = \mathbf{Q}_0.$$

In this way, the coefficients of dynamic stiffness are different from the static ones in the fact that they are determined when the system is excitated by harmonic external forces. The coefficients of dynamic stiffness (compliance) are defined as the ratio between the corresponding amplitudes of the harmonic vibrations and the external forces. These coefficients are already dependent upon the frequency of the vibrations ω. The physical meaning of these coefficients is defined analogously to Eq. (2.4).

$$r_{sj} = \frac{Q_{j0}}{x_{s0}} \quad \text{at}\, x_j \equiv 0\, (j \neq s) \quad (s, j = 1 \ldots n).$$

Therefore, the coefficients of dynamic stiffness are equal to the ratio between the amplitude of the harmonic exciting force Q_{j0} in node j and the amplitude of the vibrations of coordinates x_s under the condition that all other coordinates $s \neq j$ are rigidly fixed.

As has already been pointed out above, the inertia matrix $\mathbf{M}$ in mechanical systems is assumed to be a diagonal one. Then the following relations are obtained for the principal and collateral coefficients of static and dynamic stiffness:

$$r_{ss} = k_{ss} - \omega^2 m_s, \quad r_{sj} = k_{sj} = r_{js}.\ r_{sj} = r_{js}.$$

2.4.2 Coefficients of Dynamic Compliance

Let us now write the motion equation using the coefficients of compliance

$$x_{s0} = \sum_{j=1}^{n} e_{sj} Q_{j0} \quad (s = 1 \ldots n). \tag{2.13}$$

In matrix form

$$\mathbf{X}_0 = \mathbf{e}\mathbf{Q}_0, \quad \mathbf{e} = [e_{rj}], \; \mathbf{e} = \begin{bmatrix} e_{11} & e_{12} & \cdots & e_{1n} \\ e_{21} & e_{22} & \cdots & e_{2n} \\ \cdots & \cdots & \cdots & \cdots \\ e_{n1} & e_{n2} & \cdots & e_{nn} \end{bmatrix}. \tag{2.14}$$

Obviously $\mathbf{e} = \mathbf{R}^{-1}$.

These equations have a form that is analogous to the method of forces in civil engineering. In analogy to the static problem, the coefficients e_{rj} are called as coefficients of dynamic compliance.

Let us now determine the coefficients of dynamic compliance. To this end, let us set in Eq. (2.13) all forces equal to zero, except for Q_s. Then

$$e_{js} = \frac{x_{j0}}{Q_{s0}} \quad \text{при } Q_j = 0\,(j \neq s)\,(s, j = 1 \ldots n). \tag{2.15}$$

Therefore, the coefficient of dynamic compliance is equal to the amplitude of the harmonic vibration of the coordinates x_j, which arise at the action of the exciting force with unitary amplitude applied to node s. At the same time in all other nodes the forces are equal to zero.

The coefficient of dynamic compliance can be obtained also as a solution of the algebraic equations (2.11). Let us assume that an exciting harmonic force with amplitude Q_s has been applied to node s. Then Eq. (2.11) will take the following form:

$$\begin{bmatrix} \delta_{11}m_1\omega^2 - 1 & \delta_{12}m_2\omega^2 & \ldots & \delta_{1n}m_n\omega^2 \\ \delta_{21}m_1\omega^2 & \delta_{22}m_2\omega^2 - 1 & \ldots & \delta_{1n}m_n\omega^2 \\ \ldots & \ldots & \ldots & \ldots \\ \delta_{n1}m_1\omega^2 & \delta_{n2}m_2\omega^2 & \ldots & \delta_{nn}m_n\omega^2 - 1 \end{bmatrix} \begin{bmatrix} x_1 \\ x_2 \\ \ldots \\ x_n \end{bmatrix} = \begin{bmatrix} \delta_{1s}Q_s \\ \delta_{2s}Q_s \\ \ldots \\ \delta_{ns}Q_s \end{bmatrix}.$$

Now we define from this equation the dynamic compliances coefficients. According to (2.15) let us suppose that to the some node s the external force with unit amplitude $Q = 1$ is applied. Then we receive from this equation the following solution

$$x_j = e_{sj} = \det \begin{vmatrix} \delta_{11}m_1\omega^2 - 1 & \cdots \delta_{1j} \cdots & \delta_{1n}m_n\omega^2 \\ \delta_{21}m_1\omega^2 & \cdots \delta_{2j} \cdots & \delta_{1n}m_n\omega^2 \\ \cdots & \cdots\cdots\cdots & \cdots \\ \delta_{n1}m_1\omega^2 & \cdots \delta_{nj} \cdots & \delta_{nn}m_n\omega^2 - 1 \end{vmatrix}.$$

Let's carry out some elementary transformations in this determinant at which in row j only one nonzero term δ_{sj} remains only. After that we obtain the following relations for principal and collateral compliances

$$e_{ss} = \delta_{ss}\frac{\mathbf{\Delta}_{ss}^{(n-i)}}{\det \mathbf{e}} \qquad e_{sj} = \delta_{sj}\frac{\mathbf{\Delta}_{sj}^{(n-2)}}{\det \mathbf{e}} \tag{2.15a}$$

Here $\mathbf{\Delta}_{ss}^{(n-1)}$ is a main minor $(n-1)$x$(n-1)$ corresponding to element e_{ss} in matrix **e**, $\mathbf{\Delta}_{sj}^{(n-2)}$ is minor $(n-2)$x$(n-2)$ obtained after deletion of s and j rows and columns. This relation has a sufficiently clear physical meaning that permits one to easily obtain these coefficients. Indeed, the denominator det **e** in (2.15a) is the frequency determinant of the system. Therefore, it is determined by the multiplication of its roots, i.e., it depends on the multiplication of the squares of its natural frequencies

$$\det \mathbf{e} = \prod_{s=1}^{n-1}\left(1 - \frac{\omega^2}{\omega_s^2}\right), \tag{2.15b}$$

ω_s is the natural frequency of the system under consideration.

For the *principal dynamic compliance* e_{ss} its numerator – the minor $\mathbf{\Delta}_{ss}^{(n-i)}$ in (2.15a) – after elimination of the row s and the column s is determined. Physically, it describes a system obtained from the initial one with rigidly fixed coordinates x_s. The frequencies of such a system are antiresonance with respect to the original one. In this way, these coefficients can be obtained from (2.15a) and (2.15b) by using the coefficient of static compliance and the resonance and antiresonance natural frequencies [52, 53]

$$e_{ss} = \delta_{ss}\frac{\prod_{s=1}^{n-1}\left(1 - \frac{\omega^2}{\omega_s^{*2}}\right)}{\prod_{s=1}^{n-1}\left(1 - \frac{\omega^2}{\omega_s^2}\right)}, \tag{2.16}$$

where δ_{ss} are the coefficients of the principal static compliance, ω_s^* is the antiresonance frequencies. It follows from (2.16) that the dynamic compliances are meromorphic functions with poles at the points $\omega = \omega_s$.

For *collateral compliance* its numerator $\mathbf{\Delta}_{sj}^{(n-2)}$, corresponds to the system, obtained from initial at fixed two nodes: s and j.

Coefficients of the principal dynamic compliance for beam with masses. As an example, we will consider a weightless beam with three masses (Fig. 2.3). Let us determine the dynamic compliance e_{11}. The exciting force $Q\cos\omega t$ is applied to the left-hand mass. Stiffness of the beam is EI. The coefficients of static compliance (influence coefficients) are known [20, 30, 106]:

$$\delta_{11} = \frac{3\,l^3}{4EI}, \quad \delta_{12} = \frac{11\,l^3}{12EI}, \quad \delta_{13} = \frac{7\,l^3}{12EI}.$$

The resonance frequencies of the beam are

$$\omega_1 = 0.609\sqrt{\frac{EI}{ml^3}}, \quad \omega_2 = 2.45\sqrt{\frac{EI}{ml^3}}, \; \omega_3 = 5.2\sqrt{\frac{EI}{ml^3}}. \tag{2.17}$$

The influence coefficients for a beam with a fixation at the point of application of the load (Fig. 2.4) are

$$\delta_{22}^{(1)} = \frac{23\,l^3}{108\,EI}, \quad \delta_{23}^{(1)} = \frac{11\,l^3}{54\,EI}, \; \delta_{33}^{(1)} = \frac{8\,l^3}{27\,EI}.$$

The antiresonance frequencies are

$$\omega_1^* = 1.47\sqrt{\frac{EI}{ml^3}}, \; \omega_2^* = 4.65\sqrt{\frac{EI}{ml^3}}.$$

The dynamic compliance e_{11} for the system in Fig.2.5 is determined with the help of (2.16) as a function of the exciting frequency ω.

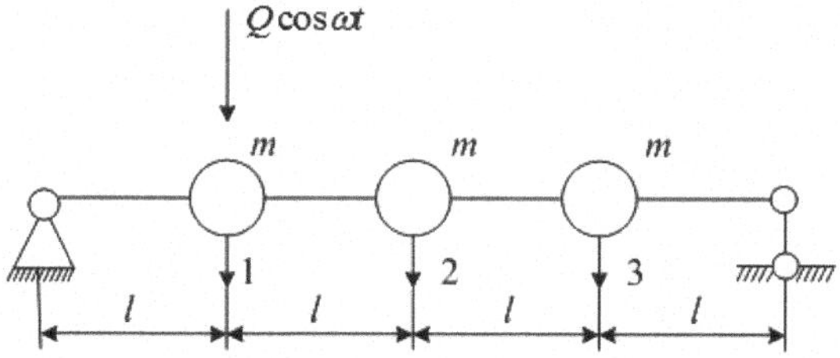

Fig. 2.3 Three-mass system with an exciting force

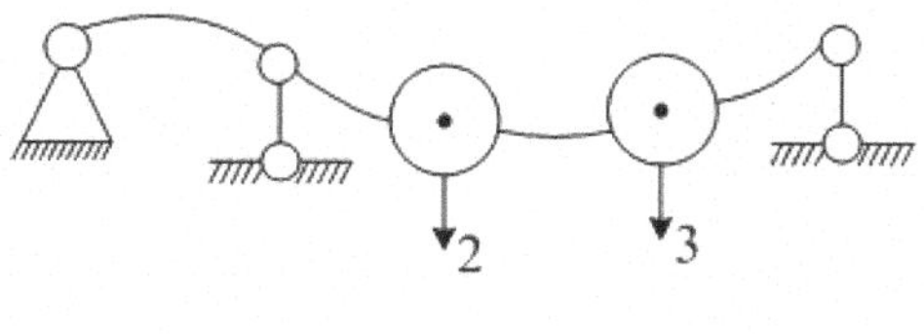

Fig. 2.4 Beam with a fixation at the point of force application

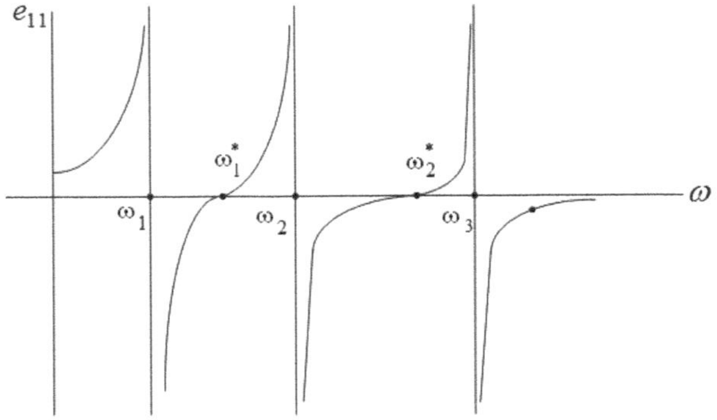

Fig. 2.5 Principal dynamic compliance e_{11} as a function of the frequency ω of the exciting force

Coefficients of the collateral dynamic compliance. When the collateral compliance e_{sj} (2.15a) is determined, the denominator remains the same and the numerator will now contain the minor $\mathbf{\Delta}_{sj}^{(n-2)}$ obtained after elimination of rows and columns s, j. It describes a system with two fixed coordinates. Then only one antiresonance frequency remains for determining $\mathbf{\Delta}_{sj}^{(n-2)}$ for the system in Fig. 2.3. Then the expression for the collateral compliance takes the form

$$e_{sj} = \delta_{sj} \frac{\mathbf{\Delta}_{sj}^{(n-2)}}{\prod\limits_{r=1}^{N} \left(1 - \frac{\omega^2}{\omega_r^2}\right)}. \tag{2.18}$$

The roots of this equation, which are determined from the equation

$$\prod (\omega) = 0,$$

can be complex conjugated. The functions determining both the principal and the collateral dynamic compliances have poles at the points corresponding to the eigenfrequencies of the system.

Let us now determine the collateral dynamic compliance for a beam (Fig. 2.3).

$$e_{31} = \delta_{31} \frac{\mathbf{\Delta}_{31}^{(n-2)}}{\prod\limits_{r=1}^{3} \left(1 - \frac{\omega^2}{\omega_r^2}\right)}, \tag{2.19}$$

where $\delta_{31} = \frac{7\,l^3}{3\,EI}$, ω_1, ω_2, ω_3 are the resonance frequencies just as in (2.17),

$$\mathbf{e}_{\text{minor}}^{31}(\omega) = 1 + \frac{\omega^2}{\omega_3^2}, \quad \omega_3^* = 3.05\sqrt{\frac{EI}{ml^3}} \text{ is the antiresonance frequencies.}$$

The dynamic compliance e_{31} obtained from (2.19) are shown in Fig. 2.6.

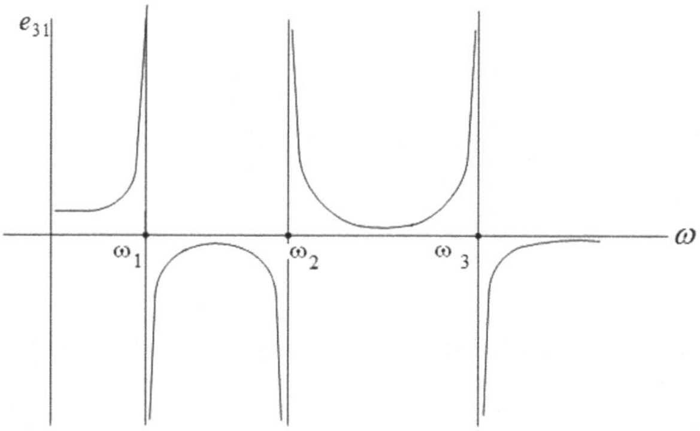

Fig. 2.6 Collateral dynamic compliance e_{31} as a function of the frequency of the exciting force ω

Therefore, the method of dynamic stiffness can be treated as a dynamic analogy of the displacements method in the statics, and the method of dynamic compliances as a dynamic analogy of the forces method in the static.

2.5 Dimension Reduction of Dynamic Compliance Matrix

One of the important advantages of the method of dynamic stiffness (compliances) is the decomposition of the system into component subsystems. When the subsystems are reintegrated, the condition for compatibility of the displacement at the interconnection point must be fulfilled. This condition means that the sum of their compliances at this point must be equal to zero: $(e'_{11} + e''_{11}) = 0$. The frequencies satisfying this equation will be the eigenfrequencies of the assembled system. The frequency equation is

$$\det(e'_{11} + e''_{11}) = 0.$$

For example, it is possible to divide a multisupport beam into two subsystems with a cross section passing through one of the supports. Then e'_{11}, e''_{11} are the dynamic compliances corresponding to the left-hand and right-hand portion of the beam. Each of these compliance matrices is of a lower order than for assembled system. So the dimension of the characteristic equation has become lower than the initial one.

This approach can also be generalized easily for systems with damping. Such a generalization is achieved by introducing a complex frequency describing the damping.

Another possibility for dimension reduction of the characteristic equation is available if additional connections are imposed on the system. If p fixing connections are imposed on the system, then zero terms emerge in the dynamic compliance matrix **e** (2.13). By means of rearrangement of the rows and columns, this matrix can be represented by a zero submatrix of the order of $(p \times p)$ in the following form:

$$\mathbf{e} = \begin{bmatrix} \mathbf{e}_{11} & \mathbf{e}_{12} \\ 0 & \mathbf{e}_{22}^{(p)} \end{bmatrix}.$$

For this equation it is possible, just as above, to apply the Gaussian elimination. If only one connection has been imposed on the system – in the x_k direction – then

$$e_{rs}^{[k]} = e_{rs} - \frac{e_{ks}e_{rk}}{e_{kk}}, \tag{2.20}$$

where the square brackets around k mean that the connection has been applied only to the generalized coordinate x_k. The dynamic compliance for a system with p additional connections is expressed by the dynamic compliance of a system with $p-1$ additional connections

$$e_{rs}^{(p)} = e_{rs}^{(p-1)} - \frac{e_{js}^{(p-1)}e_{rj}^{(p-1)}}{e_{jj}^{(p-1)}}.$$

This equation can be used for determining the dynamic compliance of a system with any number of additional connections. If the connection impedes the relative displacement of two subsystems, then (2.20) takes the form

$$\tilde{e}_{rs} = e_{rs} - \frac{e_{ks}e_{rk}}{e_{kk} + e_{kk}^{*}}. \tag{2.21}$$

Here $\tilde{e}_{rs}$ is the dynamic compliance of the connected system, e_{ks}, e_{rk}, e_{rs} are the dynamic compliance of the first subsystem, and e_{kk}^{*} is the dynamic compliance of the second subsystem.

2.6 Determining Dynamic Compliance Using Experimental Methods

When studying a new structure, it is expedient to determine the dynamic compliances of its elements. There are several experimental methods for determining the value of the dynamic compliance. For example, it is possible to measure the dynamic displacement of the elastic system caused by the action of a harmonic force that is applied at a certain point. In this case, the value of the dynamic compliance in the given frequency range of the exciting force is obtained directly. However, this method requires the specific experimental setup. In addition, for practical purposes it cannot be used for details with small sizes. That is why the following method is used for determining the dynamic compliance. The static compliance of the system and two frequency spectra are determined experimentally: a resonance and an antiresonance spectrum. In this case, the principal dynamic compliance is determined from (2.16).

This method cannot be applied directly for determining the collateral dynamic compliance. However, under certain boundary conditions, this compliance can be expressed with the help of the principal dynamic compliance. Actually, in (2.20), we have the following equation for the principal dynamic compliance at $s = r$:

$$e_{rr}^{[k]} = e_{rr} - \frac{e_{rk}^2}{e_{kk}}. \quad (2.22)$$

From this equation we can find

$$e_{rk} = \sqrt{(e_{rr} - e_{rr}^{[k]})\, e_{kk}}, \quad (2.23)$$

or

$$e_{rk} = \sqrt{(e_{kk}^{[r]} - e_{kk})\, e_{rr}}. \quad (2.23a)$$

It follows from (2.23) and (2.23a) that in order to obtain the collateral dynamic compliance e_{rk}, it is necessary to have two values of the principal dynamic compliance e_{rr}, e_{kk} and one value of the principal compliance $e_{rr}^{[k]}$ obtained by imposing on the system an additional connection along the x_k coordinate.

In this way, it is necessary to determine three static compliances $\delta_{rr}, \delta_{rk}, \delta_{kk}$ and the spectrum of the following four subsystems: the initial system, the system with an additional connection in the x_k direction, the system with an additional connection in the x_r direction, and the system with two additional connections x_k, x_r.

We will illustrate the experimental determination of the *collateral dynamic compliance* with an example of joint vibrations of the gas turbines blades with a pair-wise ring shrouding (Fig. 2.7). Two neighboring blades are connected together with a special bar passing through a hole in the input wedge of one blade and in the output wedge of another blade.

For the system in Fig. 2.7, the collateral dynamic compliance of the blades is e_{12}, and x_1, x_2 are the generalized coordinates along the bar axis (Fig. 2.8). The value e_{12} was determined from (2.21). The blade was fixed on the table of an electrodynamic vibrator, which excited the vibrations of the blade. The spectrum of the natural frequencies was determined for the two ways of blade fixation: a regular console blade, a blade with additional support in the direction of the generalized coordinate x_1, with additional support in the direction of the generalized coordinate x_2, and two additional supports in the x_3 and x_2 directions (Fig. 2.9). The static compliances $\delta_{11}, \delta_{12}, \delta_{22}$ were also determined experimentally.

Fig. 2.7 Pair-wise ring shrouding of blades

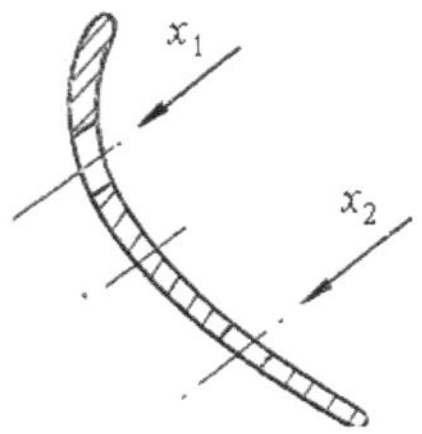

Fig. 2.8 Cross section of the blade. x_1, x_2 are generalized coordinates

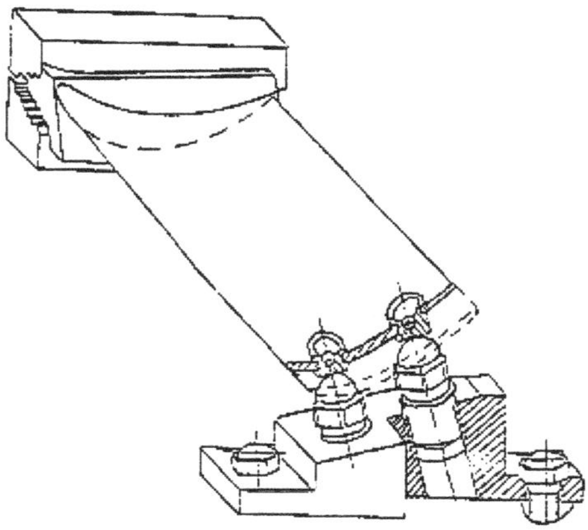

Fig. 2.9 Unit for experimental determination of dynamic compliance

In order to determine the natural frequencies of the blade in the required frequency range, one can limit the number of terms in the numerator and the denominator of Eq. (2.16) to 3 or 4. This means that it is sufficient to determine 3–4 natural frequencies of the blade under the specified boundary conditions at the locations for mounting the bars.

After determining all frequencies, the values of $e_{11}, e_{22}, e_{11}^{(2)}, e_{22}^{(1)}$ were found. From (2.23) one can find the value of the collateral dynamic compliance e_{12}. Equation (2.24) was used for control purposes. The relative value of the collateral compliance $\bar{e}_{12}$ as a function of the relative frequency of the exciting force $\bar{\omega}$ is shown in Fig. 2.10.

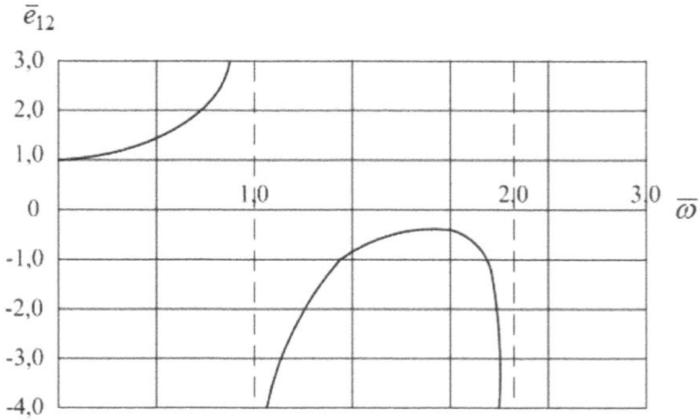

Fig. 2.10 Dependence between the relative collateral compliance of the blade $\overline{e}_{12}$ and the relative exciting frequency $\overline{\omega}$

Here $\bar{e}_{12} = \frac{e_{12}}{\delta_{12}}$, $\bar{\omega} = \frac{\omega}{100}$, e_{12} is the collateral dynamic compliance of the blade, δ_{12} is the static blade compliance, and ω is the frequency of the exciting force.

The curve in Fig. 2.10 is analogous to the collateral compliance e_{31} for a beam with three masses (Fig. 2.6).

The experimental values of the principal and the collateral compliances were used in calculating the vibrations of the blade packet.

2.7 Fundamentals of Finite-Element Method. Analytical Approaches

The finite-element method (FEM) can be used also as a method for deriving a computational mathematical model and as a method for its further numerical calculation. Usually FEM is used as a powerful tool for numerical calculations [40, 114]. However, much less attention is paid to its capabilities for obtaining interesting analytical results. Here below we pay special attention to the analytical view of FEM matrices [94, 95] and receiving analytical solutions [15].

When the FEM is used, the original system is decomposed into small finite elements. In order to determine the elastic-inertial characteristics of the finite element, an approximation of the function of the displacement (strains, etc.), expressed through the values of the coordinates on it is set up: $Z(x, y, z)$. This approximation is usually performed as an interpolation polynomial. The coefficients of the stiffness matrix $\mathbf{K} = [k_{ij}]$ for the finite element are determined from the equality of the potential energy for the finite element and the original element. In particular, for a continuous beam its potential energy is:

$$W = \sum\sum k_{ij}x_i x_j = \int (EIZ''(x))^2 \mathrm{d}x. \tag{2.24}$$

The interpolation polynomial for a beam element can be selected in the following form:

$$Z(x) = a_1x^3 + a_2x^2 + a_3x + a_4 = \tilde{a}_1H_1 + \tilde{a}_2H_2 + \tilde{a}_3H_3 + \tilde{a}_1H_4,$$

where H_i is a Hermitean function which allow to satisfying the necessary boundary conditions of beam:

$$H_1 = 1 - 3x^2/l^2 + 2x^3/l^3; \quad H_2 = x - 2x^2/l + x^3/l^2;$$
$$H_3 = 3x^2/l^2 - 2x^3/l^3; \quad H_4 = -x^2/l + x^3/l^2,$$

l is the length of the end element. Then from (2.24) it is easy to determine the respective coefficients. Therefore, FEM can be treated as a variant of the Ritz method where the coordinate functions are piecewise-polynomial.

In the case of more complex types of end elements – plane or three-dimensional ones – the interpolation functions are more complex. So, for an elastic systems described by a fourth-order differential equation, the interpolation function has the following form: $Z(x,y) = \sum\sum \beta_{ij}\, x_i\, y_j$.

The stiffness and the inertia matrices for the finite elements are available in many books [17, 94]. All necessary matrices are automatically formed in FEM software packages. The finite -element model of the shell of a nuclear reactor is presented in Fig. 2.11 as an example.

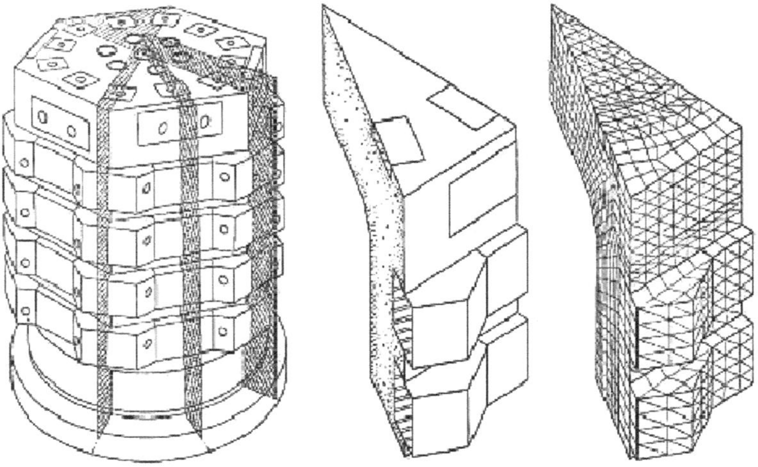

Fig. 2.11 Finite-element model of shell for a nuclear reactor

If only numeric results are needed, then the standard FEM programs can be used.

2.7.1 Stiffness Matrix for Beam Finite Element

The basic ideas of the finite elements method and its analytical capabilities will be demonstrated on an example of a beam finite element [17]. A plane beam finite element and coordinate axes at its first and the second ends are shown in Fig. 2.12a. By taking into account relations (2.6a) and (2.6b) for the static coefficients of stiffness of a beam and considering the accepted coordinate system, we find that the stiffness matrix will be as follows:

$$\mathbf{K} = \begin{bmatrix} \mathbf{K}_{11} & \mathbf{K}_{12} \\ \mathbf{K}_{21} & \mathbf{K}_{22} \end{bmatrix} = \begin{bmatrix} \frac{12EI}{l^3} & \frac{6EI}{l^2} & -\frac{12EI}{l^3} & \frac{6EI}{l^2} \\ & \frac{4EI}{l} & \frac{6EI}{l^2} & \frac{2EI}{l} \\ \textit{symmetric.} & & \frac{12EI}{l^3} & -\frac{6EI}{l^2} \\ & & & \frac{4EI}{l^3} \end{bmatrix}. \tag{2.25}$$

Block $\mathbf{K}_{11}$ (2×2) corresponds to the first end of the beam, and $\mathbf{K}_{22}$ (2×2) to the second end. The inertia matrix $\mathbf{M}$ for the finite element is determined analogously. The generalized coordinates for each end of the beam are the transversal displacements and the angular turns.

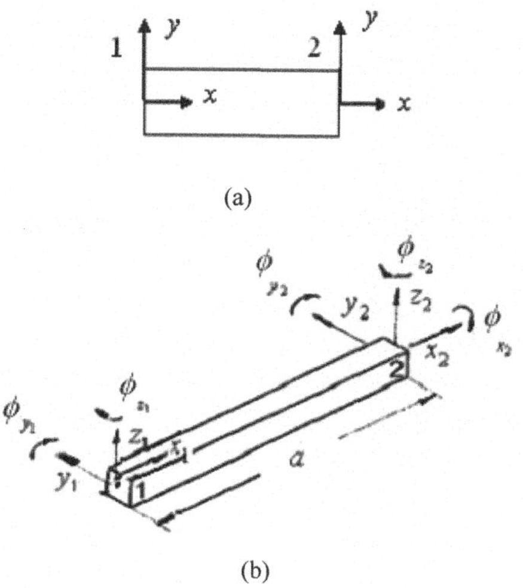

Fig. 2.12 Finite element of a beam and its local coordinate system; (**a**) two-dimensional finite element, (**b**) three-dimensional element

Let us find now the stiffness and inertia matrices for a three-dimensional finite element. Its coordinate axes at the first and second ends are selected according to Fig. 2.12b [15].

In the case of a three-dimensional beam finite element, the stiffness and inertia matrices are of the 12th order because at each beam end 1 and 2 there are six coordinate: transversal displacements in the direction of the x, y, z coordinate axes as well as turns around these axes ϕ_x, ϕ_y, ϕ_z (Fig. 2.12). This stiffness matrix is available in Appendix A.

It is possible to improve the precision of the calculations by increasing the number of finite elements. It should be noted that the finite -element model of the beam has a wavelike character unlike the Bernoulli beam (2.6), which was discussed above.

The inertia matrix **M** has an analogous structure (Appendix A), but it is very often considered a diagonal one:

$$\mathbf{M} = \text{diag}\,(m,\, m,\, m.\, I_x,\, I_y,\, I_z).$$

The stiffness matrix for the damper-type element in the (x,y) plane is

$$\mathbf{K}_{sh} = \begin{bmatrix} \mathbf{k}_{sh} & -\mathbf{k}_{sh} \\ -\mathbf{k}_{sh} & \mathbf{k}_{sh} \end{bmatrix}, \; \mathbf{k}_{sh} = \begin{bmatrix} k_x & & \\ & k_y & \\ & & k_\theta \end{bmatrix}, \tag{2.26}$$

where k_x and k_y are the damper stiffness along the coordinate axes and k_θ is the stiffness around the z axis . And the three-dimensional case is analogous.

For systems composed by several beams, there are analytical ratios for the selection of the necessary number of elements n for each beam depending on the given precision $\Delta\omega$ of the solution and the maximum frequency $\omega_{\max}$ (or the number of the normal modes) that guarantee the equivalence of the original distribution system and its finite -element modeling. (Chap. 3).

2.7.2 Stiffness Matrix for Assembled System

If the stiffness (compliance) matrix for each finite element is known, it is possible to construct a matrix for the assembled system. This can be done from the compatibility conditions at the nodes of the finite elements. Therefore, for a system consisting of several beam finite elements, its stiffness matrix is formed by summing at each node i the stiffness matrices of the elements of the respective node:

$$\hat{\mathbf{K}}_{ii} = \sum_{\mathrm{s}} \mathbf{N}_{\mathrm{s}}^{T} \mathbf{K}_{\mathrm{s}} \mathbf{N}_{\mathrm{s}},$$

where s is the number of finite elements in the i node, sub index "T" denote the transpose matrix and $\mathbf{N}_{\mathrm{s}}$ is the transformation matrix for this element which is equal

$$\mathbf{N}_{s} = \begin{bmatrix} \boldsymbol{\theta}_{\mathrm{s}} & \boldsymbol{\theta}_{\mathrm{s}}\mathbf{V}_{\mathrm{s}} \\ \mathbf{0} & \boldsymbol{\theta}_{\mathrm{s}} \end{bmatrix}. \qquad \text{(A.2), (2.27)}$$

This matrix is taking into account the shift matrix $\mathbf{V}_{\mathrm{s}}$ and the rotation matrix $\boldsymbol{\theta}_{\mathrm{s}}$ of the local coordinate system of the finite element s with respect to the general coordinates system in the node i. Matrices $\mathbf{V}_{\mathrm{s}}$ and $\boldsymbol{\theta}_{\mathrm{s}}$ can be found in Appendix A.

If there is only rotation of the local coordinate's axes $\boldsymbol{\theta}_{\mathrm{s}}$ without any shift, then the Eq. (2.27) have the more simple form:

$$\mathbf{N}_{\mathrm{s}} = \begin{bmatrix} \boldsymbol{\theta}_{\mathrm{s}} & 0 \\ \mathbf{0} & \boldsymbol{\theta}_{\mathrm{s}} \end{bmatrix}. \qquad (2.27a)$$

This transformation matrix $\mathbf{N}_{\mathrm{s}}$ describes only the rotation of the coordinate axes. Further on, we will use mainly this transformation matrix.

Therefore, the finite-element stiffness matrix for the assembled system can be presented in block form (Appendix A):

$$\hat{\mathbf{K}} = \begin{bmatrix} \hat{\mathbf{K}}_{11} & \hat{\mathbf{K}}_{12} & \cdots & \hat{\mathbf{K}}_{1n} \\ \hat{\mathbf{K}}_{21} & \hat{\mathbf{K}}_{22} & \cdots & \hat{\mathbf{K}}_{2n} \\ \cdots & \cdots & \cdots & \cdots \\ \hat{\mathbf{K}}_{n1} & \hat{\mathbf{K}}_{n2} & \cdots & \hat{\mathbf{K}}_{nn} \end{bmatrix}, \qquad (2.28)$$

where n is the number of nodes. Node i is described by the diagonal block $\mathbf{K}_{\mathrm{ii}}$; the off-diagonal blocks $\hat{\mathbf{K}}_{\mathbf{ij}} = \mathbf{N}_{\mathbf{i}}^{\mathbf{tr}}\mathbf{K}_{\mathbf{ij}}\mathbf{N}_{\mathbf{j}}$ describe the elastic interaction between the i and j nodes. One important and convenient feature is the fact that the structure of this matrix reflects the structure of the system.

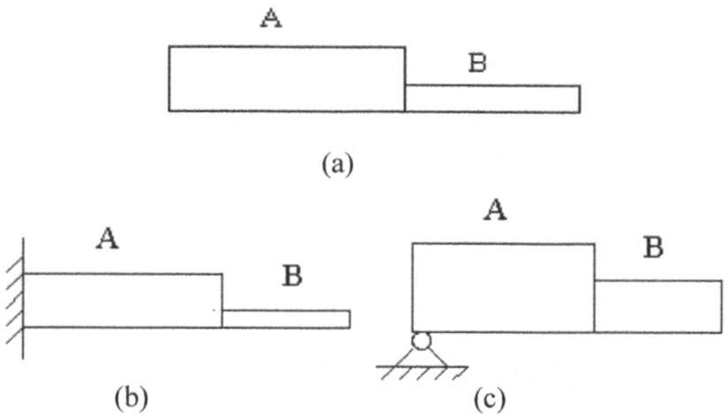

Fig. 2.13 Two identically positioned finite elements at different boundary conditions

Definition. Let us call partial a subsystem that is obtained from the general system by fixing rigidly all others subsystems (nodes).

Therefore, we can consider node i as a partial subsystem that is described by matrix $\mathbf{K}_{ii}$.

If the system represents a no ramified unidirectional structure, then in this case sufficiently simple rules for deriving the stiffness (and inertial) matrix are obtained. So when two identically positioned beam finite elements are *combined* in some node, the stiffness matrix is formed by summing in this node the blocks of the appropriate stiffness matrices.

For example, let us consider the simplest system consisting of two identically positioned beam elements A, B (Fig. 2.13a). The transformation matrix $\mathbf{N} = \mathbf{E}$ ($\mathbf{E}$ is the identity matrix). The stiffness matrix has the following form (at free ends):

$$\hat{\mathbf{K}} = \begin{bmatrix} \mathbf{K}_{11}^{\mathrm{A}} & \mathbf{K}_{12}^{\mathrm{A}} & \\ & \mathbf{K}_{22}^{\mathrm{A}} + \mathbf{K}_{11}^{\mathrm{B}} & \mathbf{K}_{12}^{\mathrm{B}} \\ \textit{symmetric.} & & \mathbf{K}_{22}^{\mathrm{B}} \end{bmatrix}. \tag{2.29}$$

If the elements are connected at an angle, then it is necessary to take into account the rotation matrix $\boldsymbol{\theta}$ in the process of their summing.

In the case of complex ramified structures, the FEM matrix is more complex and it is necessary to use Eqs. (2.27) and (2.28).

Finally we receive the equation for forced and free (at $\mathbf{Q} = 0$) vibrations for FEM model

$$\mathbf{M}\ddot{\mathbf{x}} + \mathbf{K}\mathbf{x} + \mathbf{B}\dot{\mathbf{x}} = \mathbf{Q}e^{i\omega t}.$$

2.7.3 Boundary Conditions and Various Ways of Subsystems Connecting

Taking into account the boundary conditions when deriving a stiffness matrix is a very simple operation. To this purpose, it is necessary to eliminate from the matrix the rows and columns that correspond to the rigidly fixed coordinates. Thus, after elimination of the respective rows and columns, the stiffness matrix for the beam system in Fig. 2.13b with its first end fixed will be [unlike (2.29)]

$$\hat{\mathbf{K}} = \begin{bmatrix} \mathbf{K}_{22}^{\mathrm{A}} + \mathbf{K}_{11}^{\mathrm{B}} & \mathbf{K}_{12}^{\mathrm{B}} \\ (\mathbf{K}_{12}^{\mathrm{B}})^T & \mathbf{K}_{22}^{\mathrm{B}} \end{bmatrix}.$$

If the first end of beam A is hinge-supported (Fig. 2.13c), then it is necessary to eliminate in block $\mathbf{K}_{11}^{\mathrm{A}}$ the rows and the columns corresponding to the transversal displacements along the respective axes and to leave the angular displacements. The same is true for the inertia matrix.

These rules are applicable also to the combining of subsystems for different ways of their interconnections.

Subsystems connecting with a damper-type elastic element. Let us consider a system consisting of two subsystems connected with a damper with stiffness $\mathbf{k}_{\mathrm{sh}}$. The stiffness matrix $\mathbf{K}$ of the connected system is

$$\mathbf{K} = \begin{bmatrix} \mathbf{K}^{\mathrm{A}} + \mathbf{k}_{sh} & -\mathbf{k}_{sh} \\ -\mathbf{k}_{sh} & \mathbf{K}^{\mathrm{B}} + \mathbf{k}_{sh} \end{bmatrix}.$$

The diagonal blocks of the stiffness matrix describe as usually the partial subsystems.

2.8 Decomposition Methods Taking into Account Weak Interactions Between Subsystems

The dynamic analysis of the vibration flows in mechanical structures with working mechanisms is effective only when the system is treated as a whole and the interactions between its subsystems and components are taken into account. The computational models of such systems (as a rule, they are finite-element models) have a hierarchical structure and contain hundreds and thousands of degrees of freedom. Therefore, considerable difficulties arise both in their simulation and in their theoretical and numerical analysis. So the following dilemma appears already at the stage of computational model deriving: Should one use a more universal model that is difficult to analyze, or a simpler one but with a shorter confidence interval. Increasing the accuracy of the numerical solution for universal model is not very efficient due both to the approximation of the model itself and the nearness of parameters in the model.

When studying large hierarchical systems of any nature, the most powerful and natural approaches are the decomposition ones. These approaches make it possible to analyze the assembled systems separately and then take into account the interactions between these subsystems. It is most efficient both in the analysis and the calculation of large systems.

It should be pointed out that the arbitrary decomposition of a mathematical model is inefficient. Taking into account the connections between subsystems substantially changes the dynamics of the system. The solution for separate subsystems differs considerably from those for an assembled system. Therefore, such approaches are necessary that take into account the effect of dynamic interactions in the system.

It was found that the principal feature of the dynamic behavior of large systems is the occurrence of weak interactions between the subsystems. The weak interaction between the subsystems is not an special case; on the contrary, it is a general physical essence of such systems. As theoretical analysis and calculation experience have shown, weak interactions, caused by various physical factors, are virtually always present in systems with a hierarchical structure. In particular they arise at large discrepancy of stiffness parameters; small asymmetry, wide range of natural frequencies [9]. Indeed these factors are usual for large systems. So the weak interactions between some subsystems are a common system feature.

The vibrations in the weak-connected subsystems take place practically independently from each other. The behavior of the assembled system is determined by strongly coupled variables and in particular by pass-through parameters: at vibrations, they concentrate the system energy. Decomposing the system in such a way takes into account the principal systemic processes, and therefore it is quite natural for large systems. It permits one to specify a limited number of dominant variables (or subsystems) that determine the dynamics of the system. At the numerical level, this permits one to carry out the significant reduction of the computational model. With this approach, the decomposition as a physical system and its mathematical model coincide; this makes it possible to achieve a deep understanding of the dynamic processes in large systems due to revealing of dominant variables.

2.8.1 Coefficients of Dynamic Interactions

Initially it is necessary to introduce strict analytical criteria for dynamic interactions. These criteria have to be dimensionless coefficients that determine the interactions that are typical for mechanical systems [10].

The concept of connection for two-mass system was presented for the first time in [76, 104]. Its equations are:

$$m_1\ddot{x}_1 + (k_{10} + k_{12})x_1 - k_{12}x_2 = 0$$
$$m_2\ddot{x}_2 + (k_{20} + k_{12})x_2 - k_{12}x_1 = 0.$$

The coefficient of the connection between them was determined as follows:

$$\gamma_{12} = \frac{k_{12}}{\sqrt{(k_{10} + k_{12})\,(k_{10} + k_{12})}}.$$

However, for multidimensional systems, these coefficients require essential generalization. Indeed, for example, the values of k_{ij} for the FE-models are defined only by the FEM mesh and do not determine the intrinsic essence of the system.

Let us now define the interaction coefficients for a multidimensional system. The equations of the vibrations for a system with many subsystems can be written in the form of a composite block matrix

$$\mathbf{D} = \mathbf{K} - \Lambda\mathbf{M} = \begin{bmatrix} [\mathbf{K} - \Lambda\mathbf{M}]_{11} & [\mathbf{K} - \Lambda\mathbf{M}]_{12} & \cdots & [\mathbf{K} - \Lambda\mathbf{M}]_{1N} \\ \cdots & \cdots & \cdots & \cdots \\ [\mathbf{K} - \Lambda\mathbf{M}]_{N1} & [\mathbf{K} - \Lambda\mathbf{M}]_{N2} & \cdots & [\mathbf{K} - \Lambda\mathbf{M}]_{NN} \end{bmatrix}, \tag{2.30}$$

where $\mathbf{K}_{ii}$ and $\mathbf{M}_{ii}$ are the stiffness and inertia matrices for the i-th partial subsystem, $\mathbf{K}_{ij}$ and $\mathbf{M}_{ij}$ are the stiffness and inertia matrices describing the connections between subsystems i and j, N is the total number of subsystems, n_i is the dimensionality of subsystem i, $\mathbf{\Lambda}_i = diag[(\omega_i^{(p)})^2]$ is the diagonal matrix of the eigenvalues for subsystem i, and $(\omega_i^{(p)})^2$ is the quadrate of the p-th natural frequency. The damping factor can be taken into account by setting up $\mathbf{K}$ as a complex matrix. Let us recall that the partial subsystem is the subsystem i that has been obtained in the case when all other subsystems $i \neq j$ are rigidly fixed.

Definition. Subsystems i and j, (for example 1 and 2), are called weakly coupled if their normal modes (in a dimensionless form) can be presented in the following way:

$$\boldsymbol{h}_1 = \begin{bmatrix} \boldsymbol{h}_{11} \\ \varepsilon\boldsymbol{h}_{12} \end{bmatrix}, \quad \boldsymbol{h}_2 = \begin{bmatrix} \varepsilon\boldsymbol{h}_{21} \\ \boldsymbol{h}_{22} \end{bmatrix},$$

where $\mathbf{h}_i$ is the matrix column of the normal modes for the first and second partial subsystems. These expressions mean that the ratio of the components of the natural modes related to another subsystem is very small and is equal to $\varepsilon \ll 1$.

The dimensionless coefficients of dynamic interaction that are typical for mechanical systems are the following [9]:

- *Energy coefficients*, which determine the ratio between the work performed by the forces of the connections between the subsystems and the energy accumulated in the partial subsystems at the given natural modes

$$\alpha_{ij}^{pr} = \frac{(\mathbf{h}_i^p)^T \mathbf{K}_{ij} \mathbf{h}_j^r}{\left(\omega_i^{(p)} \omega_j^{(r)}\right)}, \; p = 1 \ldots n_i, \; r = 1 \ldots n_j, \tag{2.31}$$

where $\mathbf{h}_i^p$ is the p-th natural mode for subsystem i.

- *Spectral coefficients*, which depend on the detuning of the frequencies in the subsystems

$$s_{ij}^{pr} = \frac{\alpha_{ij}^{pr}}{(1 - \omega_i^{(p)}/\omega_j^{(r)})}. \tag{2.32}$$

These coefficients are small for weakly coupled subsystems.

$$\alpha_{ij}^{pr} < \varepsilon, \; s_{ij}^{pr} < \varepsilon.$$

From a physical point of view, these conditions mean that the energy in the connections between the subsystems is much less than the energy in the subsystems themselves. At vibrations, a weak energy exchange (ε-order) takes place between them. It follows from Eqs. (2.31) and (2.32) that in order to determine the dynamic interactions coefficients it is enough to know only the solutions for the subsystems but not for the assembled system.

After the interaction coefficients have been determined, the following matrices are formed from them: $\boldsymbol{\alpha}_{ij} = [\alpha_{ij}^{pr}]$, $\mathbf{s}_{ij} = [s_{ij}^{pr}]$. These matrices are characterized by the dynamic interaction between subsystems i and j at all modes. Then the criteria for the weak dynamic interactions are:

$$\left\|\boldsymbol{\alpha}_{ij}\right\| \leq \varepsilon, \; \left\|\mathbf{s}_{ij}\right\| \leq \varepsilon. \tag{2.33}$$

Here $\left\|\boldsymbol{\alpha}_{ij}\right\|$, $\left\|\mathbf{s}_{ij}\right\|$ are the norms of the given matrices.

Let us now show that matrix $\mathbf{D}$ for weakly coupled subsystems can be presented in a form containing the small parameter ε. To do that, it is necessary first to represent the matrix in a dimensionless form. This can be achieved by means of the matrix transformation

$$\mathbf{D}^* = \left(\boldsymbol{\Lambda}^{-1/2}\mathbf{H}\right)^T \mathbf{D} \left(\boldsymbol{\Lambda}^{-1/2}\mathbf{H}\right) = \left[\alpha_{ij}^{pr}\right] = \boldsymbol{\alpha}_{ij} =$$
$$\begin{bmatrix} 1-\Lambda/\omega_1^2 & \alpha_{12} & \cdots & \alpha_{1n} \\ \alpha_{21} & 1-\Lambda/\omega_2^2 & \cdots & \alpha_{2n} \\ \cdots & \cdots & \cdots & \cdots \\ \alpha_{n1} & \alpha_{n1} & \cdots & 1-\Lambda/\omega_n^2 \end{bmatrix},$$

where $\mathbf{H}$ is the matrix built from the vector columns of the natural forms for all partial subsystems. It can be seen from this that if the weak interaction conditions (2.33) are fulfilled, ε-blocks appear in matrix $\mathbf{D}^*$. The structure of the matrix that describes the weak-connected systems consists of diagonal blocks with an ϵ-envelope around them that is, as a rule, rather rare (Fig. 2.14):

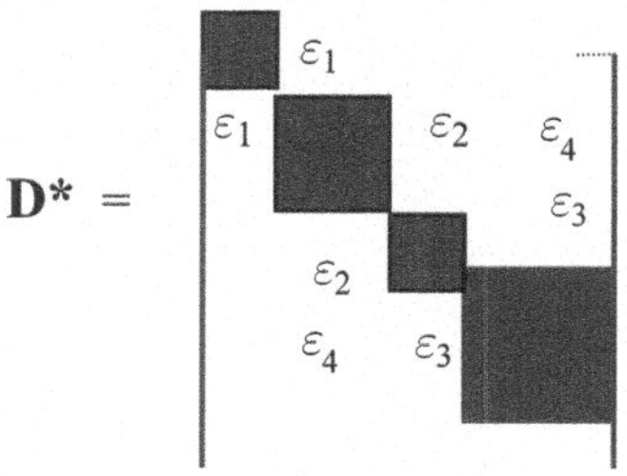

Fig. 2.14 The structure of matrix for weak-coupled system

Table 2.1 Decomposition ways depending on physical nature of weak interactions and their position

Weak interaction types	$[\alpha_{ij}] < \varepsilon$, $[\mathbf{s}_{ij}] \neq \varepsilon$	$[\alpha_{ij}] < \varepsilon$, $[\mathbf{s}_{ij}] < \varepsilon$	$[\alpha_{ij}] \cong 1$, $[\mathbf{s}_{ij}] < \varepsilon$
Dynamic matrix structure	$\begin{bmatrix} \mathbf{D}_{11} & \varepsilon\mathbf{D}_{12} \\ \varepsilon\mathbf{D}_{21} & \mathbf{D}_{22} \end{bmatrix}$	$\begin{bmatrix} \mathbf{D}_{11} & \varepsilon\mathbf{D}_{12} \\ \varepsilon\mathbf{D}_{21} & \varepsilon^2\mathbf{D}_{22} \end{bmatrix}$	$\begin{bmatrix} \mathbf{D}_{11} & \mathbf{D}_{12} \\ \mathbf{D}_{21} & \mathbf{K}_{11} - \Lambda\varepsilon\mathbf{M}_{22} \end{bmatrix}$
Decomposition type	Partition on independent substructures $\mathbf{D}_{11}$ and $\mathbf{D}_{22}$	Elimination of substructure with dynamic matrix $\mathbf{D}_{22}$	The aggregation of elements
Conditions of occurrence	Wide scatter of stiffness parameters; small asymmetry	The large range of natural frequencies in system	Wide scatter between partial frequencies

The position of the ε-blocks at various physical peculiarities of the system, is shown in Table 2.1. The ways of decomposition, will be different depending on the position of the weak interactions.

2.8.2 Decomposition by Partition into Independent Substructures

Let us discuss now decomposition by partition into independent subsystems. This approach will be used frequently later on.

The vibrations in weakly coupled subsystems take place practically in independently; their properties given below are very useful:

- the natural frequencies of a weakly coupled system are very close – to within ε – to the natural frequencies of the untied partial subsystems;
- the normal modes at assembling weakly coupled subsystems vary a little in the absence of close frequencies in the subsystems. However, the normal modes corresponding to close frequencies vary abruptly, which points to an exchange of energy in resonance modes.

At $\varepsilon = 0$, a generating system is obtained. It describes partial subsystems that are not interconnected. It can be considered as a comparison system.

The solution of the assembled system can be looked for in the form of a convergent series in powers of ε [9, 10].

In the nonresonance case, the following conditions are fulfilled:

$$\omega_i^{(p)} \neq \omega_j^{(r)}, \quad \| \boldsymbol{\alpha}_{ij} \|, \| \mathbf{s}_{ij} \| \leq \varepsilon,$$

The solution has the following view (in matrix form)

$$\mathbf{\Lambda} = \mathbf{\Lambda}_0 + \varepsilon^2 \mathbf{\Lambda}^{(1)} + \ldots, \qquad \mathbf{H} = \mathbf{H}_0 + \varepsilon \mathbf{H}_0 \mathbf{S} + \ldots, \tag{2.34}$$

$\mathbf{\Lambda}_0$, $\mathbf{H}_0$ are matrices of the eigenvalues and natural modes of the generating (not interconnected) system at $\varepsilon = 0$.

In the resonance case, the following conditions are fulfilled:

$$(1 - \omega_i^{(p)}/\omega_j^{(r)})\varepsilon, \parallel \boldsymbol{\alpha}_{ij} \parallel < \varepsilon, \parallel \mathbf{s}_{ij} \parallel > 1/\varepsilon).$$

The solution, relating close frequencies has the following form:

$$\omega_{i,j} = \omega_{i0} \pm \alpha_{is} \Delta\xi + o(\varepsilon^2), \; \mathbf{h}_i = \mathbf{h}_i \pm \mathbf{h}_j + \varepsilon s_{ij} \mathbf{h}_i + o(\varepsilon), \tag{2.34a}$$

where the index "0" refers to generating (not interconnected) partial system. The series (2.34), (2.34a) are geometric progressions and because of that they quickly converge.

It can be seen that the partial subsystems describe most closely the behavior of the assembled system.

As an example, we will analyze the dynamic interactions in the subsystems of a gas-turbine engine (GTE) [2]. This engine has three main subsystems (Fig. 2.15): low-pressure rotor (LPR) (1), high-pressure rotor (HPR) (2), and a casing (C) (3).

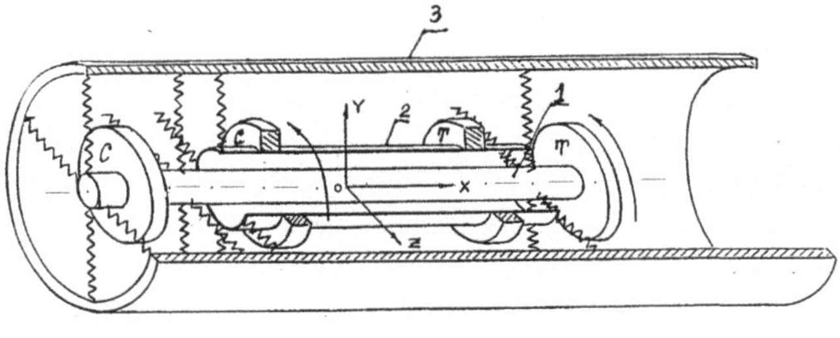

Fig. 2.15 Finite-element model of a gas-turbine engine and its principal subsystems

As shown by the theoretical and numerical analysis, there is a weak dynamic interaction in the low-frequency range between the casing of the engine and the rotors. In addition, weak dynamic interactions take place in the low-pressure rotor between the compressor and the turbine in this frequency range. This is confirmed by Table 2.2. The natural frequencies of the not interconnected partial subsystems (LPR, HPR, C) and the natural frequencies of the assembled system of the gas-turbine engine are close to each other.

The presence of weak interactions in the engineering structures plays a double role. On the one hand, this is very useful because they increase the vibroisolation of the subsystems and impede the transmission of vibrations in the structure. On the other hand, however, they complicate the monitoring of the defects because of the weak transmission of the signal inside the construction.

Table 2.2 Natural frequencies of the partial subsystems of the GTE and the assembled system

RLP	27,8	66	134,9	221,1	954		
RHP	142	241	339,5	1251			
Case	132	217	2189				
Coupled system.	27	53	65,13	72,45	137	157	238

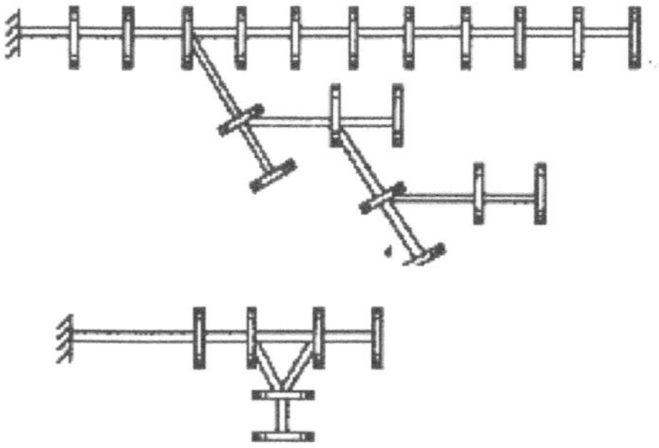

Fig. 2.16 Gradual aggregation of the computational model of a lathe shaft line

2.8.3 Other Decomposition Methods

Decomposition by integration of the elements. This decomposition approach can be used, for example, when the system contains high-frequency partial elements or subsystems. The aggregation is performed by an elimination of the high-frequency elements and redistribution their stiffness-inertia characteristic to other elements. In this case, only dominant elements determining its energy remain in the mathematical model. Then the model is considerably reduced. Such approach to non-ramified systems was proposed for the first time in [98]. In [9], it has been generalized for the general case of linear systems. The example in Fig. 2.16 presents an aggregation process for a ramified computational model of a lathe shaft line. Fig. 2.16 shows the gradual elimination of high-frequency elements. This approach allows one to reveal the elements determining the vibration process and at the same time reduce the degrees of freedom from 21 to 6. The accuracy of the natural frequency calculation when using the minimal model is approximately 4%.

Elimination of substructures. This method solves the problem of obtaining a minimal mathematical model in the given frequency band ($0 - \omega_{Max}$). The method is based on eliminating the natural modes with weak influence in this frequency range.

This method was used for dynamic analysis of an antenna structure consisting of three subsystems connected by rolling element bearings [9]. The finite-element model initially had almost 5000 degrees of freedom. In the course of constructing a reduction model, natural modes for each subsystem were found in a frequency range of up to 60 Hz. The minimal model in the 0–30 Hz frequency range, which was obtained after the assembling of subsystems, has only 24 degrees of freedom with an accuracy of numerical calculation of the frequencies of about 10%.

The mathematical models obtained as a result of above decomposition methods have a general energy level close to that of the initial system with an accuracy of up to ε.

2.8.4 Coefficients of Weak Interaction and Criteria for Ill – Conditions of Matrices

The coefficients of dynamic interactions (2.31) and (2.32) appear to be sufficiently universal characteristics for large systems. Indeed, we will show that they are in agreement with the following criteria:

1. The criteria of ill- conditions of the matrix at forced vibrations [8]
$$\rho(\mathrm{D}) = \omega_{\min}/\omega_{\max} << 1$$
coincides with condition $\alpha_{ij}^{pr} > \varepsilon$.
2. The criteria of ill-conditions of the natural modes [111]
$$\rho(\mathrm{h}) = \min(1 - \omega_i^{(p)}/\omega_j^{(r)})/\max(1 - \omega)_i^{(p)}/\omega j^{(r)}) > 1/\varepsilon$$
coincides with coefficient $s_{ij}^{pr} > \varepsilon$.
3. The sensitivity criteria for the natural modes of the i-th subsystem at varying the parameters of the j-th subsystem is characteristic by coefficient α_{ij}^{pr}.
4. The criteria for equivalence of various mathematical models between each other can be expressed in the following way:
$$\| \alpha_{ij} \|_1 - \| \alpha_{ij} \|_2 \leq \varepsilon, \qquad \| s_{ij} \|_1 - \| s_{ij} \|_2 \leq \varepsilon.$$

These conditions are very important because the choice a computational model of a system is ambiguous and does not represent a formalized procedure. The choice of the computational model often depends on the researcher's intuition and experience. That is why it is very important to have criteria for the equivalence of various computational models between one another.

On the basis of the above-obtained results, it is possible to formulate the following general rule for modeling of systems. The coefficients of dynamic interactions (2.31) and (2.32) depend on the frequency ω. Therefore, different types of weak interactions can arise in a system depending on the frequency range. Consequently, the dynamic properties and, consequently, the position of the ε-blocks in matrix $\mathbf{D}^*$ will change. Therefore, it is reasonable to have a frequency hierarchy of the models in the case of a large system. Actually, in this case in each frequency range we will obtain a reduced model containing only the dominant parameters. In addition, we have a well-conditioned dynamic matrix for the numerical procedure.

Therefore, decomposition into weak-coupled subsystems is not only a numerical approach. This process can give answers to a number of problems already at the engineering study: in solving the problems related to subsystem arrangement, the necessary accuracy in manufacturing of the individual elements, and further on in experimental studies – for determining the points for optimal positioning of the sensors.

Part I
Systems with Lumped Parameters

Chapter 3
Vibrations of Regular Systems with Periodic Structure

3.1 Introduction: Some Specific Features of Mechanical Systems

Systems with periodic structure consisting of repeated elements very much frequently are used in mechanical constructions. They also is designated as systems with translational symmetry. A very effective method in the study of vibrations for such systems is the use of a dispersion equation [25, 69, 89, 97]. This equation determining the dependence of the vibrations frequency on the wavelength: $\omega = F(\kappa)$, where κ is wave number (or vector), $\kappa = \dfrac{2\pi}{L}$, L is wavelength, ω here and further in this section denotes the current vibrations frequency.

As a rule, the low- and medium-frequency ranges are of interest in the mechanical engineering. In this case, it is necessary to take into account the effect of the boundary conditions and the connection points. Since dispersion equation does not depend on the boundary conditions, it makes it possible to analyze the specific characteristics of the structure itself. In order to obtain a complete solution, it is necessary to take into account also the boundary conditions. These conditions are given by the frequency equation. The simultaneous solution of these equations allows one to find a complete solution of the eigenvalue problem for a given system.

Generally speaking, the frequency equation is a complex enough transcendental equation. However, in certain cases its roots are equidistant [24, 109], as we shall see below, so that the simultaneous solution of these equations can be easily obtained graphically.

The possibility for analysis of the dispersion equation and the frequency equation has clear advantages. This permits one to separate the properties depending on the type of structure from the properties that are determined by the fixing conditions. This fact is particularly important for the analysis and further optimization of mechanical systems.

Generally speaking, the wave properties of an arbitrary system with discrete-distributed parameters are difficult to describe. However, it appears that the use of discretization with the help of the finite-element method (FEM) permits one to do that because the FEM model – as will be shown below – has a wave nature. The use of the dispersion equation also permits one to determine the limits of

L.Y. Banakh, M.L. Kempner, *Vibrations of Mechanical Systems with Regular Structure*, Foundations of Engineering Mechanics, DOI 10.1007/978-3-642-03126-7_3,

the equivalence area of the continual and discrete models of a system and also to perform an evaluation of the coincidence for these models.

In Sect. 3.3, a method for deriving dispersion equations using the FEM is discussed. Sect. 3.4 contains the results of a study of a two-layer laminas with sparsely positioned laminar ribbings. The corresponding dispersion surfaces and normal modes for different frequency ranges are provided.

Section 3.6 is devoted to the analysis of beam systems. The use of the dispersion equation for finite-elements models of beam systems makes it possible to obtain an exact analytical solution for such models and compare it with the solution for systems with distributed parameters. An error estimation as a function of the finite-element mesh has also been performed. A table is given that allows one to determine the necessary number of finite elements as a function of the required numerical accuracy and frequency range (or for a given number of oscillation modes).

3.2 Wave Approach at Vibrations of Mechanical Systems with Periodic Structure

The main results which may be received from the analysis of dispersion equations will be demonstrated using as an example a simple mechanical system with periodic structure. As a very simple system, we will discuss the torsional vibrations of a weightless shaft with identical disks mounted on it (Fig. 3.1). Suppose at first that this system is endless.

The equation for the rth disk is

$$J\omega^2\varphi_r + k\left(\varphi_{r-1} + \varphi_{r+1} - 2\varphi_r\right) = 0, \tag{3.1}$$

where k is the stiffness of the shaft between the disks and J is the inertia moment of disk.

This equation is a finite-difference equation. It is known [22, 42] that an n- order homogeneous linear finite-difference equation has n linearly independent particular solutions. These solutions form a fundamental system. The general solution of the homogeneous equation is a sum of the linearly independent particular solutions and contains n arbitrary constants. The particular solution (3.1) in the absence of multiple roots has the form

$$\varphi_r = Ce^{i(\omega t - \mu r)}. \tag{3.2}$$

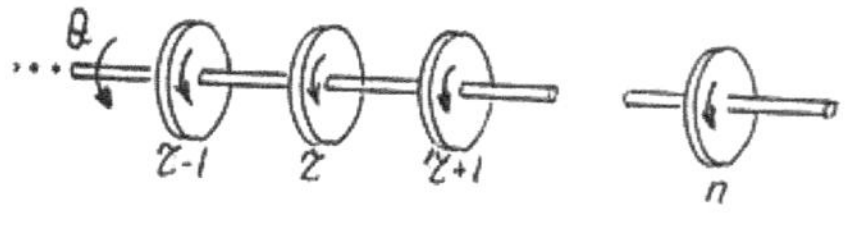

Fig. 3.1 System with periodic structure

This solution can be treated as a wave. In general it is the traveling wave. and in this case ω in (3.2) means the current frequency, but at standing wave ω coincides with the natural frequency.

The solution (3.2) one may represent also in a view

$$\varphi_r = Ce^{i(\omega t - \kappa x_r)} = Ce^{i(\omega t - \kappa ar)}, \tag{3.2a}$$

where: a is the distance between the neighboring masses in the system.[1]

It is seen that parameter μ characterizes the phase shift in the transition from element r to element $r+1$ and it is connected with wave number κ with help of relation

$$\mu = \kappa a.$$

The note. Solution (3.2) can be treated also as a projection operator of a group with translational symmetry (Chap. 4).

From Eqs. (3.1) and (3.2) we find the dispersion equation that relates the oscillation frequency to the wave parameter μ:

$$\cos\mu = 1 - \frac{J\omega^2}{2k}. \tag{3.3}$$

As follows from Eq. (3.3), the frequency ω is a periodic function of the μ. Let us consider the interval $-\pi \le \mu \le \pi$ (Fig. 3.2).

It can be seen from Eq. (3.3) that the harmonic solution exists only when $\omega < \omega_0$, where $\cos\mu < 1$. In this case Eq. (3.3) has two real solutions: $\pm\mu$, and then solution (3.2) takes the form

$$\varphi_r = C_1 \text{Cos}\mu r + C_2 \text{Sin}\mu r. \tag{3.4}$$

From here it follows that the wavelength is equal to $L_w = \frac{2\pi}{\mu}a = \frac{2\pi}{\kappa}$.

In order to determine the natural frequencies, it is necessary to solve – along with the dispersion equation – the frequency equation. The frequency equation can be obtained by setting the boundary conditions at $r = 0$, $r = n+1$. In this way,

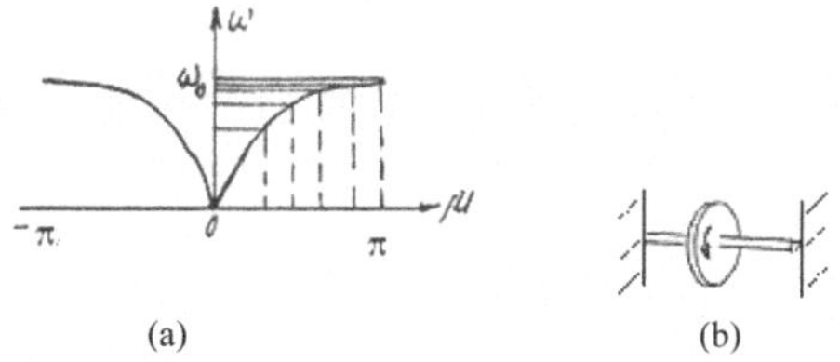

Fig. 3.2 **a, b** Dispersion curve (**a**), partial subsystem corresponding to frequency ω_0 (**b**)

[1]Usually the wave number is designated as k, but to not confuse it to stiffness coefficient here we have designated it through κ

in particular, we obtain from Eq. (3.4) for the system Fig 3.1 with fixed ends the known equation

$$\mu\,(n+1) = \pi r. \tag{3.5}$$

The analysis of Eqs. (3.3), (3.4), and (3.5) and Fig. 3.2 permit one to obtain important information about the system properties, namely:

---

(1) *The existence of a pass-band range of the harmonic signal:*

$$0 < \omega < \omega_0,\ \omega_0^2 = \frac{2k}{J}.$$

Therefore the system in Fig. 3.1 can be considered as a high-frequency filter. Really at $\omega > \omega_0$, the solution (3.2) will not be harmonic since here μ is a complex quantity: $\mu = ib$, and one may see from Eq. (3.2) the vibrations amplitudes damped in proportion to e^{br}.

(2) *The existence of frequency condensation points at $n \to \infty$.*

The simultaneous graphical solution of (3.5) and (3.3) (Fig. 3.2a) shows the existence of a point of frequency condensation – a limit frequency ω_0. This frequency corresponds to a minimal wavelength equal to $\mu_{\lim} = \pi n/(n+1) \to \pi$, where all neighboring masses vibrate in an anti phase. The limit frequency ω_0 is determined from the dispersion equation. Therefore, it does not depend on the boundary conditions. This corresponds to the known fact that the effect of the boundary conditions on shortwave vibrations is weak. In this case, the periodic structure can be considered endless. One may see that the maximal frequency ω_0 is defined as frequency for partial one-mass subsystem with fixed ends (Fig. 3.2b) and so may be easy determined.

(3) *Comparing the accuracy of the discrete and continuous models with distributed parameters for a system with a periodic structure.*

A continuous model can be obtained from the discrete model at $n \to \infty$ (if such a limit exists). In our case, such a limit is possible, and the equation describing this model is the wave equation:

$$\frac{\partial^2\varphi}{\partial t^2} - v^2\frac{\partial^2\varphi}{\partial x^2} = 0,$$

where $v = \pm\omega/\kappa$-is phase wave velocity. Its dispersion equation:

$$\omega = \pm v\kappa = \pm\mu v/a \tag{3.6}$$

This equation shows a linear dependence between ω and μ. Equation (3.6) coincides with the dispersion equation (3.3) for a discrete system if $\mu << 1$. Really (3.3) in the low-frequency range will be $\mu = \pm\omega/\omega_0$, ω_0 is partial frequency for elementary subsystem Fig. 3.2b. It illustrates Fig. 3.3.

Fig. 3.3 Comparison of dispersion curves for discrete model (1) and with distributed parameters one (2)

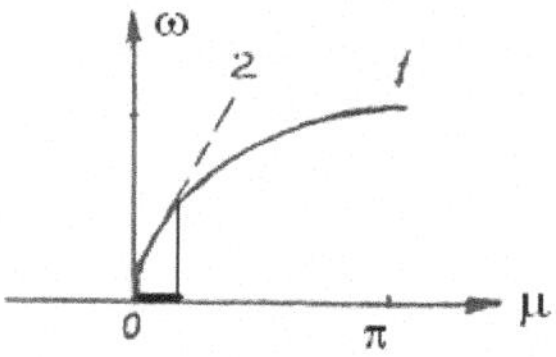

By comparing (3.6) and (3.3) with each other, it is possible to evaluate the differences between the continuous and discrete models and to find the area where their equations differ little from each other (Fig.3.3).

(4) *Determining limit wavelength for various normal modes.*

Let us complicate the model in Fig. 3.1 by assuming that disks of two kinds alternate in the system: J_1 and J_2 (Fig. 3.4a). Then the corresponding dispersion equation takes the form [89]

$$\omega^2 = k\left[\frac{1}{J_1}+\frac{1}{J_2}\pm\sqrt{\left(\frac{1}{J_1}+\frac{1}{J_2}\right)^2-\frac{4\sin^2\mu}{J_1J_2}}\right]^{1/2} \qquad (3.7)$$

The dispersion curve (3.7) is shown in Fig. 3.4b.

The bottom branch is acoustic; it corresponds to low-frequency vibrations.

The top branch is optical; it corresponds to high-frequency vibrations. It can be seen from Fig. 3.4b that the dispersion curve is multivalued along μ: two frequencies correspond to the same wavelength – ω_b for the bottom branch and ω_t for the top branch. Therefore, there are two normal modes (Fig. 3.5). These modes

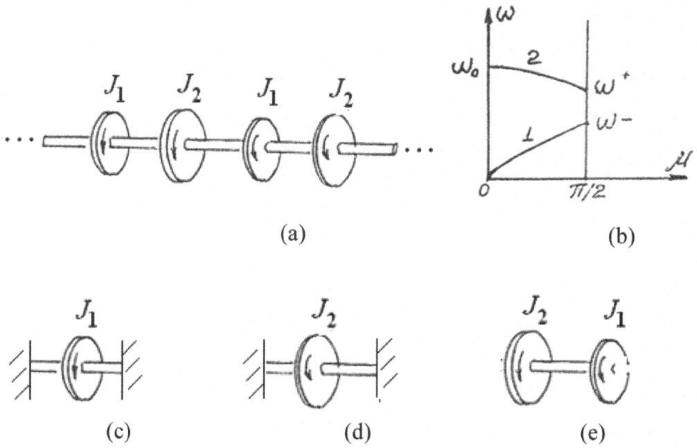

Fig. 3.4a–e A system with two kinds of alternating disks (**a**), its dispersion curve (**b**), partial subsystem corresponding to ω_- (**c**), partial subsystem corresponding to ω_+ (**d**), partial subsystem corresponding to ω_0 (**e**)

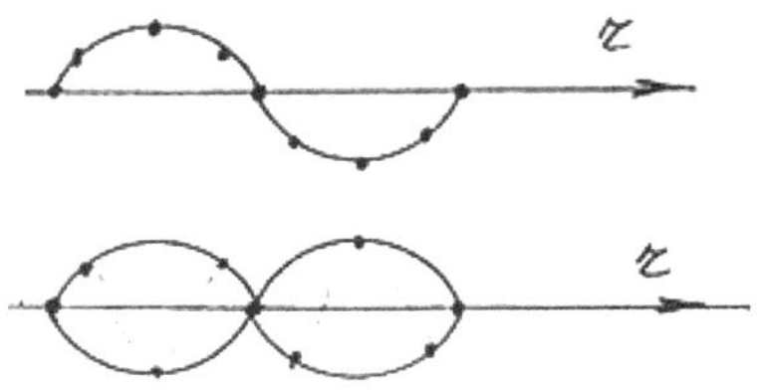

Fig. 3.5 Normal modes

have the same wavelength, but they describe in-phase vibrations (ω_b) and antiphase vibrations (ω_t) of the neighboring masses. In this way, the dispersion equation permits one to establish the phase relationships between the neighboring elements, i.e., the configuration of the normal modes and the frequency limits for the different normal modes.

The interval $\omega_- \div \omega_+$ (Fig. 3.4b) corresponds to the nonpass band of the vibrations (the filter zone); In this area, the amplitude of the vibrations are damped out exponentially. ω_0 is the limit frequency for harmonic vibrations

$$\omega_- = \sqrt{\frac{2k}{J_1}}, \quad \omega_+ = \sqrt{\frac{2k}{J_2}}, \quad \omega_0 = \sqrt{2k\left(\frac{1}{J_1} + \frac{1}{J_2}\right)};$$

Figure 3.4c–e shows, that these important frequencies correspond to very simple partial subsystems. Therefore one can easily to define them.

The nonpass band occurs at $\mu = \dfrac{\pi}{2}$. Let us use this fact for determining the boundaries of the low frequencies in a periodic structure of the type shown in Fig. 3.1. If on the system presented in Fig. 3.4 the differences between J_1 and J_2 will decrease, then it will describe in the limit the system shown in Fig. 3.1. Combining in pairs the neighboring masses in the system in Fig. 3.1, we can transform the curve in Fig. 3.2 analogously to Fig. 3.4 (Fig. 3.6).

It follows from here that the point $\mu \geq \pi/2$ in Fig. 3.6a corresponds to the antiphase vibrations of the neighboring masses. Continuing this process further, combining the q neighboring masses (i.e., forming quasiparticles or super elements), we will find that the quasiparticles of q masses are in a phase at $\mu \leq 2\pi/q$, etc. (Fig. 3.6b).

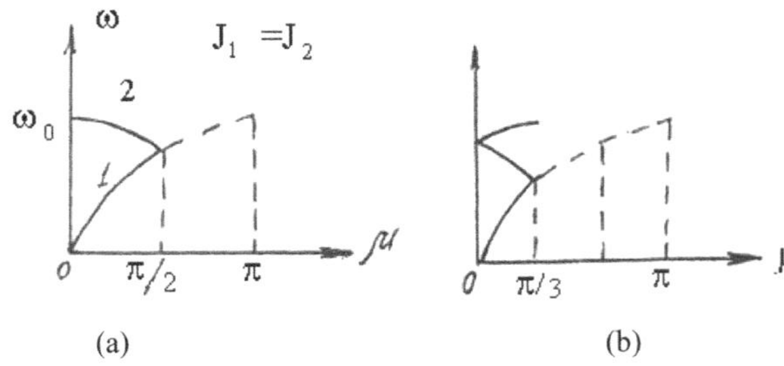

Fig. 3.6a, b Determining the limit for the lowest frequency

The second frequency is obtained at antiphase vibrations of $n/2$ neighboring masses and arise at q=$n/2$, i.e., at $\mu_2 \leq 4\pi/n$. This is the upper limit $\omega_{1\,\max}$ of the first low frequency (at any boundary conditions); and always $\omega_2 \geq \omega_{1\,\max}$. The difference in the wavelengths $\Delta\mu$ for the neighboring oscillation modes is equal to $2\pi/n$, which explains the equidistant spectrum of the wave numbers.

Let us note that the nonpass bands correspond to the Brilluen zone [25].

Although the provided data have been obtained for rather simple oscillation systems, the basic properties will also be correct for more complex continuous-discrete systems, which are considered below.

3.3 Vibrations of Frames with Periodic Structure

As a rule, mechanical systems with a periodic structure are continuous-discrete systems. Usually they are of periodic structures where the inhomogeneities are concentrated in discrete regular points (they are called geometric grating). They can be one-, two-, and three-dimensional periodic systems. The dispersion equation for such systems has certain differences compared to those discussed in Sect. 3.2. First, the dispersion equation can contain at the same time terms with $\cos\mu$ ($\sin\mu$) and $ch\mu$ ($sh\mu$), and, therefore, its periodicity is not preserved. Second, there will be no limit frequency in systems having the distributed parameters. Actually, in the case of sufficiently high frequencies, the discrete inclusions will remain fixed, performing the role of a rigid support for the elements with distributed parameters. In such a case, there are an unlimited number of natural frequencies.

Various methods for constructing the dispersion equation for continuous-discrete systems can be proposed. In [6], a method for constructing the dispersion equation for grids with beam elements with the help of dynamic rigidities is proposed. Below we will use FEM models that permit one to describe a sufficiently broad class of systems.

3.3.1 Combining Finite Elements Method and Dispersion Equation

In this case, if the structure consists of sufficiently complex subsystems or elements for which it is difficult to find analytically a dynamic stiffness matrix, then the apparently universal FEM is used.

With the help of the FEM, it is possible to construct a dispersion equation for a large class of systems, and at that in analytical form. In this case, the order of the equation coincides with the dimensionality of one elementary cell.

The general form of the matrix describing the vibrations of a periodic chain system formed with the help of the FEM has the following form:

$$\mathbf{D} = \mathbf{K} - \lambda\mathbf{M} = \begin{bmatrix} [\mathbf{D}_{11}] & [\mathbf{D}_{12}] & & \\ & [\mathbf{D}_{11}] & [\mathbf{D}_{12}] & \\ & & \ddots & [\mathbf{D}_{11}] \end{bmatrix}. \tag{3.8}$$

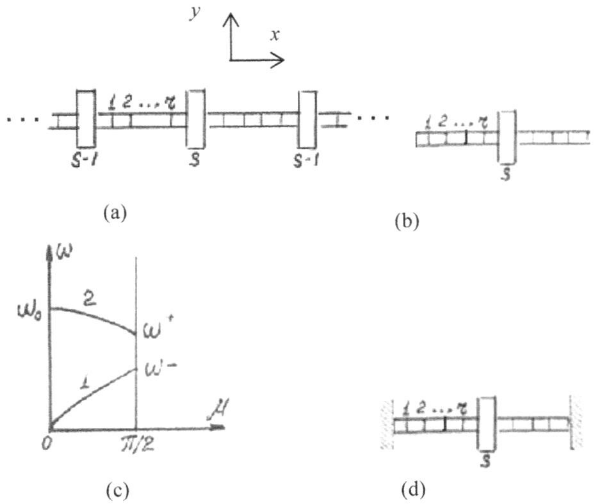

Fig. 3.7a–d The finite-element model of a beam (**a**), elementary subsystem (**b**), dispersion curves (**c**), elementary subsystem with fixed ends (**d**)

$[\mathbf{D}_{ii}] = [\mathbf{D}_{11}] = \mathbf{K}_{11}$-$\lambda\mathbf{M}_{11}$ is a matrix describing the vibrations of node i, $[\mathbf{D}_{12}]$ is a connection matrix between the i th and the $i+1$ nodes.

The general form of the dispersion equation is

$$\left|\mathbf{D}_{21}e^{-\mu} + \mathbf{D}_{11} + \mathbf{D}_{12}e^{\mu}\right| = 0. \qquad (3.9)$$

In the case of ramified systems, the equation has a more complex form and is discussed in Sect. 3.4.

As a simple example, let us find the dispersion equation for a beam with identical rigid bodies mounted on it; they have the same mass M_c, and inertia moment J_c (Fig. 3.7a).

Let us consider the bending vibrations in the (*x*, *y*) plane. Each node i has two coordinates – the displacement and the rotation angle: (y, θ_z). Let us partition the section of the beam between the bodies into r finite elements. For the sake of simplicity, we will assume that the mass of the beam is concentrated in the partition nodes and then the inertia matrix of the beam (J_6, m_6) is diagonal. Let us write down a fragment of the stiffness and inertia matrices for some section s with a disk:

$$\underset{(6r\times 6r)}{\mathbf{D}_{11}} = \begin{array}{c} \begin{matrix} 1 & 2 & 3 & \ldots & r \end{matrix} \\ \begin{bmatrix} \mathbf{K}_{11} & \mathbf{K}_{12} & & & \\ \mathbf{K}_{21} & \mathbf{K}_{11} & \mathbf{K}_{12} & & \\ & \mathbf{K}_{21} & & \ddots & \mathbf{K}_{11} \end{bmatrix} \end{array} - \omega^2\,[\mathbf{M}],$$

$\underset{(6r\times 6r)}{[\mathbf{M}]} = \mathrm{diag}\,[M_c + m_b, J_c + J_b, m_b, J_b, \ldots J_b]$, $\mathbf{K}_{ij}$ are the blocks of the stiffness matrix for beam with r finite elements.

The dispersion equation according to Eq. (3.9) is

$$f(\omega,\mu) = \begin{bmatrix} 1 & 2 & 3\ \ldots\ r \\ \mathbf{K}_{11} & \mathbf{K}_{12} + \mathbf{K}_{21}e^{-\mu} & \\ & \mathbf{K}_{11} & \mathbf{K}_{12}\ \ldots \\ \hline \mathbf{K}_{21} + \mathbf{K}_{12}e^{\mu} & \ldots & \mathbf{K}_{11} \end{bmatrix} - \omega^2\mathbf{M},$$

$$\det f(\omega,\mu) = 0.$$

Opening this equation, we shall receive a quadratic equation about $ch\mu$ and of a degree $2r$ about ω^2. Hence, the equation will have $2r$ branches of ω for one value μ. Each optical branch, similarly Fig. 3.5, will correspond to the internal vibrations modes of partial subsystems Fig. 3.2b.

Taking into account the actual expressions for $\mathbf{K}_{11}$, K_{12} (Chap. 2), we can easily write dispersion equation.

In the low-frequency range, we can set the number of nodes in Fig 3.7a at $r = 1$. Then (3.9) can be written in the following form

$$\begin{vmatrix} \left(24\frac{EJ}{l^3} - (M_c + m_b)\omega^2\right) - 24\frac{EJ}{l^3}\cos\mu_r & 12\frac{EJ}{l^2}\sin\mu_r \\ -12\frac{EJ}{l^2}\sin\mu_r & \left(8\frac{EJ}{l} - (J_c + J_b)\omega^2\right) - 4\frac{EJ}{l}\cos\mu_r \end{vmatrix} = 0 \tag{3.10}$$

l -length of finite element. This equation is of the fourth order with respect to e^{μ}, even on μ ; therefore, it has four solutions for μ:

$$\mu^{1,2} = \pm\mu_1,\ \mu^{3,4} = \pm\mu_2.$$

To obtain a periodic solution satisfying the boundary conditions, it is necessary that the quantity μ_1 be real and μ_2 be purely imaginary, i.e., $\mu_2 = i\,|\mu_2|$. It has been taken into account that $ch\mu_2 = \cos|\mu_2|$, $sh\mu_2 = i\sin|\mu_2| = \sin\mu_2$.

Developing (3.10), we obtain a quadratic equation with respect to $\cos\mu(ch\mu)$ and an equation of the order of 4 with respect to ω. This dispersion equation is of second order; consequently it has two branches (Fig. 3.7c) which is completely similar to Fig. 3.4c. One can receive the nonpass band ($\omega_- \;-\; \omega_+$) which arise if $\cos\mu \geq 1$. Then at $\mu = \pi/2$ we have from Eq. (3.10) the following equation for definition of ω_- , ω_+

$$\begin{vmatrix} \left(-(M_c + m_b)\omega^2\right) & 12\frac{EJ}{l^2} \\ -12\frac{EJ}{l^2} & \left(4\frac{EJ}{l} - (J_c + J_b)\omega^2\right) \end{vmatrix} = 0. \tag{3.11}$$

One can see that equation (3.11) describes the subsystem Fig 3.7b but with rigidly fixed ends (Fig. 3.7d). Therefore the nonpass range corresponds to the frequencies of elementary subsystem with fixed ends.

The limit frequency for system under consideration one can receive at $\mu = 0$. From Eq. (3.10) we find that

$$\omega_0 = \frac{4EJ/l}{(J_c + J_b)},$$

that is the partial frequency for torsion vibrations of elementary subsystem.

Evidently similar results one may receive and for elementary subsystem with distributed parameters.

3.3.2 Vibrations of Grate Frames

The use of dispersion equations which have the low dimensionality becomes particularly efficient in the study of plane and spatial grating frames. Actually, the number of DOF in such systems is rather large – on the order of n^2 (or n^3) (n is the number of finite-elements nodes); But the dispersion equation dimensionality is equal to the number of DOF for single elementary section.

Let us study the vertical bending vibrations of a plane frame (Fig. 3.8) consisting of beam elements.

The node now has a double numbering – for the longitudinal row (t) and for the transversal one (j). The following coordinates are included in the stiffness matrices of the beams: for longitudinal beams (z, θ_y), for transversal beams (z, θ_x). In this way, each node has three coordinates (z, θ_x, θ_y). It is supposed that the all masses of

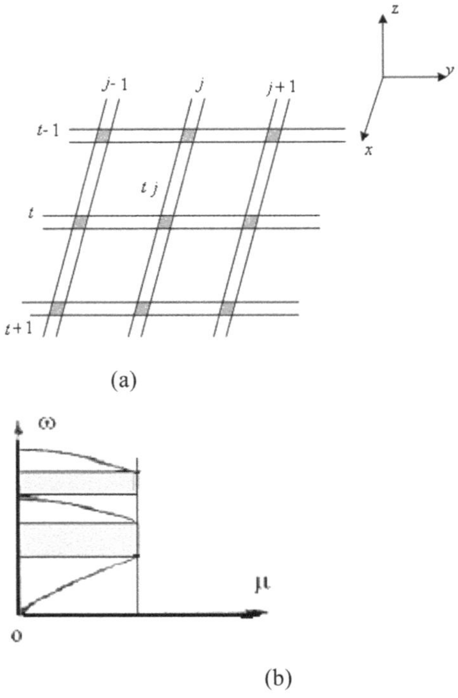

Fig. 3.8a, b A grate frame (**a**) and its dispersion curves (**b**)

finite elements for longitudinal beams and for transversal beams M_1, M_2 are located in intersections nodes, their inertia moments about axis x, y are J_{ix}, J_{iy} $(i = 1,2)$; between the nodes there is an one beam finite element. Now let us write the equation for the displacement of node tj:

$$\mathbf{K}_{11}\mathbf{Z}_{tj} + \mathbf{K}_{21}^{(2)}\mathbf{Z}_{tj+1} + \mathbf{K}_{12}^{(1)}\mathbf{Z}_{t+1,j} + \mathbf{K}_{12}^{(2)}\mathbf{Z}_{tj-1} + \mathbf{K}_{21}^{(1)}\mathbf{Z}_{t-1j} - \mathbf{M}\mathbf{Z}_{tj}\cdot\omega^2 = 0. \quad (3.12)$$

In this equation, the upper index "1" refers to longitudinal beams and index "2" to transversal ones.

By treating (3.12) as finite-difference equations, we will search for its solution in the form

$$\mathbf{Z}_{t+p,\,j+r} = \mathbf{X}_t e^{i(\mu p+\omega t)}\mathbf{Y}_j e^{i(\nu r+\omega t)},$$

where μ and ν are the wave parameters in the longitudinal and the transversal directions, respectively. Then from Eq. (3.12) we obtain the dispersion equation

$$\left|(\mathbf{K}_{11}\text{-}\mathbf{M}\omega^2) + \left(\mathbf{K}_{21}^{(2)}e^{\nu} + \mathbf{K}_{12}^{(2)}e^{-\nu}\right) + \left(\mathbf{K}_{12}^{(1)}e^{\mu} + \mathbf{K}_{21}^{(1)}e^{-\mu}\right)\right| = 0.$$

By taking into account the actual expressions for the stiffness matrix of the beams (Chap. 2, Appendix A), we obtain the following dispersion equation:

$$\det\begin{bmatrix} a_{11} & -6a_{12}sh\nu & -6a_{13}sh\mu \\ 6a_{12}sh\nu & a_{22} & 0 \\ 6a_{13}sh\mu & 0 & a_{33} \end{bmatrix} = 0, \quad (3.13)$$

$$a_{11} = \frac{24\mathrm{E}\,J_1}{l_1^3} + \frac{24\mathrm{E}\,J_2}{l_2^3} - (M_1 + M_2)\,\omega^2;$$

$$a_{22} = \frac{8\mathrm{E}}{l_2} - (J_{1x} + J_{2y})\,\omega^2;$$

$$a_{33} = \frac{8\mathrm{E}\,J}{l_r} - (J_{1y} + J_{2x})\,\omega^2.$$

J_1, J_2 are the inertia moments of the beams.

Analyzing Eq. (3.13), we can see that the view of the dispersion curve and the width of the nonpass areas depend on the direction of the wave propagation.

In the case of propagation of the wave in the horizontal direction $(\nu = 0)$, we obtain from Eq. (3.13) that the dispersion equation is decomposed into two independent ones:

$$a_{22} + 2G_1(1 - ch\mu) - \frac{4EJ_2}{l_2} = 0; \quad \det\begin{bmatrix} a_{11} & -\frac{12EJ_1}{l_1^2}sh\mu \\ \frac{12EJ_1}{l_1^2}sh\mu & a_{33} - \frac{4EJ}{l_1}ch\mu \end{bmatrix} = 0. \quad (3.14)$$

The first equation corresponds to the rotation vibrations around the x axis. The second equation is analogues to Eq. (3.10): therefore it describes the vibrations of a single beam with a periodically concentrated load (M, J'), and with an increased inertia moment. Therefore a frame in this case is equivalent to a discrete-loaded beam. The nonpass area is determined from Eq. (3.14). If we consider a large number of finite elements between the frame nodes, then the quantity of nonpass areas and optical branches increases. The dispersion curves are characterized by more and more narrowed nonpass bands (on Fig. 3.8b are shaded).

In the case of wave propagation under an angle of 45° ($\mu = \nu$), we obtain from Eq. (3.14) for a frame with identical horizontal and transversal beams

$$\left(a_{22} + 2G_1(1 - ch\mu) - \frac{4EJ_2}{l_2}\right)\begin{vmatrix} \frac{48EJ}{l^3}(1 - ch\mu) - M\omega^2 & -\sqrt{2}\frac{12EJ_1}{l_1^2}sh\mu \\ -\sqrt{2}\frac{12EJ_1}{l_1^2}sh\mu & \frac{8EJ_1}{l_1} - 2G_2(1 - ch\mu) - \frac{4EJ_1}{l_1}ch\mu - J_1'\omega^2 \end{vmatrix} = 0.$$

The nonpass bands at $M = J = 0$ disappear and the dispersion equation coincides with the equations for a non-loaded beam.

3.4 Dynamic Properties of Laminar Systems with Sparsely Positioned Laminar Ribbing

In the modern mechanical engineering and in civil building, systems consisting of ribbed laminas and shells are widely used. In particular, they include various kinds of honeycomb structures. The use of these elements makes the structures light, economical, and sufficiently rigid. However, the analysis of the dynamics of such structures becomes extremely difficult and can yield unexpected results. Indeed, the computational models of such systems is rather discrete and contains a large number of degrees of freedom. The continualization and transition to continuous computational models is a rather widely used method. Thus the discrete-periodic system is replaced by a smooth one with some equivalent characteristics [3, 36, 70]. As a rule, such replacements are obtained on the basis of a number of hypotheses, which requires an evaluation of their areas of application. A similar continualization is usually applied to systems with stiffening ribs positioned relatively close to one another. However, in the case of sparsely positioned and rather thick ribbings, continualization is possible only within a very limited range of parameters change, and often it is not possible at all. In addition, the problem of selecting the best structural positioning of just the ribbing elements emerges during the design process. All this requires a dynamic analysis of the discrete model.

In this respect, it is expedient to use a dispersion equation for such systems, which permits one to take into account analytically the discreteness as well as the periodicity of the structure in the space. By comparing the obtained dispersion equation with the known exact equations of a continuous medium for laminas and shells, it is possible to determine areas in which continualization is possible and to find equivalent parameters for systems with distributed parameters. On the other hand, the dispersion equation permits one to analyze such properties as nonpass bands, the points of frequencies concentration that are available only in the discrete systems. Below it is shown in such a way that two-layer laminas, with sparsely positioned and thick laminar stiffness in practically cannot be replaced by an orthotropic lamina; such a replacement is possible only in a very narrow frequency range.

3.4.1 Dispersion Equation for Ribbed Laminar Systems: Conditions for Possibility of Continualization

Let us consider the laminar system presented in Fig. 3.9 consisting of a two-layer lamina with laminar stiffness ribs [1]. In the same way as above, we will use a dispersion equation.

By using the stiffness and inertia FEM matrices for laminar elements [94], we write the equilibrium equation for the *SR* th node (Fig. 3.10).

$$\sum_{r=S-1}^{S+1}\sum_{j=R-1}^{R+1}\left(\mathbf{K}_{rj}\mathbf{Z}_{rj}+\mathbf{M}_{rj}\ddot{\mathbf{Z}}_{rj}\right)+\sum_{r=S'-1}^{S'+1}\left(\mathbf{K}_{rR'}\mathbf{Z}_{rR'}+\mathbf{M}_{rR'}\ddot{\mathbf{Z}}_{rR'}\right)=0, \tag{3.15}$$

where $\mathbf{K}_{rj}$ and $\mathbf{M}_{rj}$ are the blocks of the stiffness and inertia matrices of the laminar elements connecting the *SR* th node with its neighboring nodes. Analogously, for the $S'\ R'$ th node

$$\sum_{r=S'-1}^{S'+1}\sum_{j=R'-1}^{R'+1}\left(\mathbf{K}_{rj}^{S'R'}Z_{rj}+\mathbf{M}_{rj}^{S'R'}\ddot{\mathbf{Z}}_{rj}\right)+\sum_{r=S-1}^{S+1}\left(\mathbf{K}_{rj}^{S'R'}Z_{rR}+\mathbf{M}_{rR}^{S'R'}\ddot{\mathbf{Z}}_{rR}\right)=0. \tag{3.16}$$

For the continuous system with distributed parameters the solution in the form of a travelling wave will be

$$\overline{\mathbf{z}}=\overline{\mathbf{C}}e^{i(\sigma_1 x+\sigma_2 y+\omega t)}, \tag{3.17}$$

where $\boldsymbol{\sigma}$ is the wave vector, σ_1 and σ_2 are the projections of the wave vector on the x and y axes.

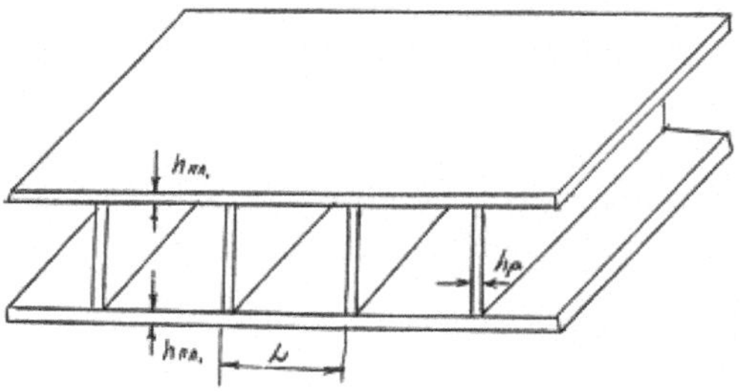

Fig. 3.9 A lamina with sparsely positioned laminar stiffening ribbings

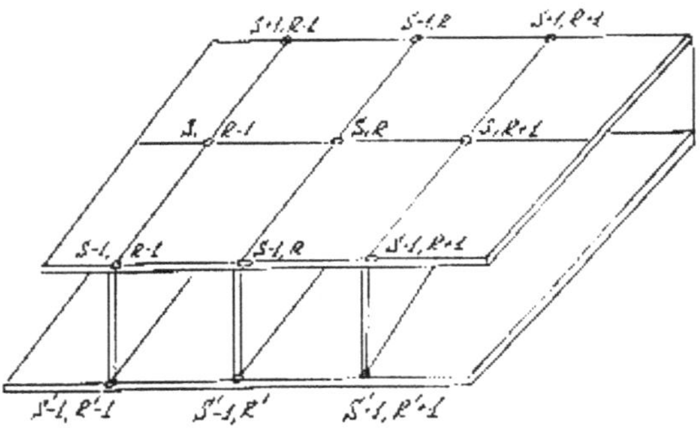

Fig.3.10 Numbering of the nodes

In the case of discrete the system, the solution of the Eqs. (3.15) and (3.16) analogues to (3.17) will have the form

$$\mathbf{z} = \mathbf{C}e^{i(\mu R+\nu S+\omega t)}$$

where $\mu = \frac{\sigma_1 x}{R}$, $\nu = \frac{\sigma_2 y}{S}$, $\mathbf{C} = \left\{C_x, C_y, C_z, C_\varphi, C_\psi, C_\theta, \dots\ C'_\theta\right\}$.

In the general case, the projections of the wave vector can be complex

$$\mu = a_1 + ib_1;\ \ \nu = a_2 + ib_2.$$

If $b_1 = 0$ and $b_2 = 0$, then the wave is homogeneous; when $a_1 = 0$ and $a_2 = 0$, the wave is no homogeneous (damped). The frequency range in which the wave is homogeneous is a pass band range; otherwise it is a nonpass band range (filter zone).

Substituting solution (3.17) in (3.15) and (3.16) and dividing into Z_{SR} and $Z_{S'R'}$, we obtain the equilibrium equation in a finite-difference form.

The structure of the studied system is symmetric with respect to the horizontal axis. Taking this property into account, we perform the coordinate transformation $\mathbf{Z}^* = \mathbf{P_2 Z}$ for symmetric and skew-symmetric modes, where

$$\mathbf{P}_2 = \left[\begin{array}{c|c} \mathbf{E} & \text{-}\mathbf{E} \\ \hline \mathbf{E} & \mathbf{E} \end{array}\right], \quad \begin{bmatrix} \mathbf{Z}_1^* \\ \mathbf{Z}_2^* \end{bmatrix} = \begin{bmatrix} \mathbf{E} & \mathbf{E} \\ \mathbf{E} & -\mathbf{E} \end{bmatrix} \begin{bmatrix} \mathbf{Z}_1 \\ \mathbf{Z}_2 \end{bmatrix},$$

E is the identity matrix. Coordinates $\mathbf{Z_1}$ and $\mathbf{Z_2}$ correspond to the top and bottom laminas, respectively. Coordinates $\mathbf{Z^*}_1$ and $\mathbf{Z^*}_2$ describe the symmetric and skew-symmetric variables.

As a result, we obtain two systems of equations corresponding to the symmetric and the skew-symmetric modes. Equating to zero the determinant of obtained equations, we receive the dispersion equation. For the analysis of the dispersion equations, it is convenient to use dispersion surfaces. These surfaces describe the bending, longitudinal, transversal-shifting, and torsion modes (waves).

The all calculations in this section were fulfilled by Dr. Sc. R.S. Akhmetkhanov

Fig. 3.11a, b. Dispersion surfaces (**a**) and symmetric modes (**b**)

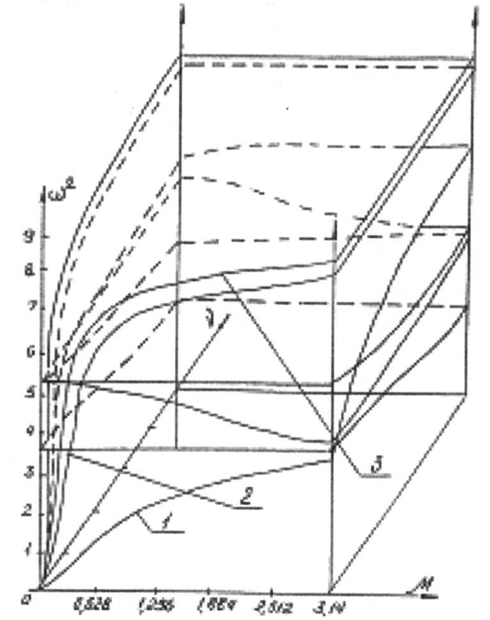

(a)

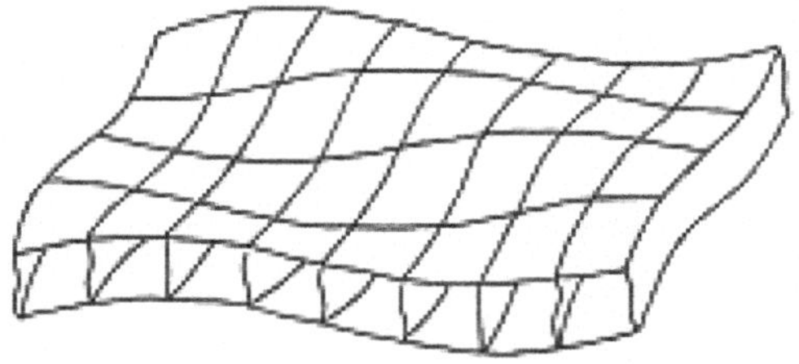

(b)

Figures 3.11a, b and 3.12a, b show the dispersion surfaces and the vibration modes for a lamina in the case where a wave propagates under an angle of $\pi/4$. The system has the following parameters: outside lamina thickness 5 mm, ribbing thickness 1 mm, distance between the ribbings equal to distance between laminas 1000 mm. In this case, the ribbings are indeed sparsely positioned.

Dispersion surfaces have characteristic features. In the case of symmetric vibrations, these surfaces, which describe the translational motions of nodes, begin from the zero point. This point corresponds to the motion of the system as a free body at $\mu = \nu = 0$. But the skew-symmetric modes start with a frequency ω^*. There are no skew-symmetric vibrations in the range from 0 to ω^*. This is easy to understand because they can only arise when there are skew-symmetric vibrations of the ribbings. It is obvious that such vibrations have higher frequencies.

The location of the surfaces permits one to determine which modes correspond to the low-frequency portion of the spectrum. The dispersion surface describing the

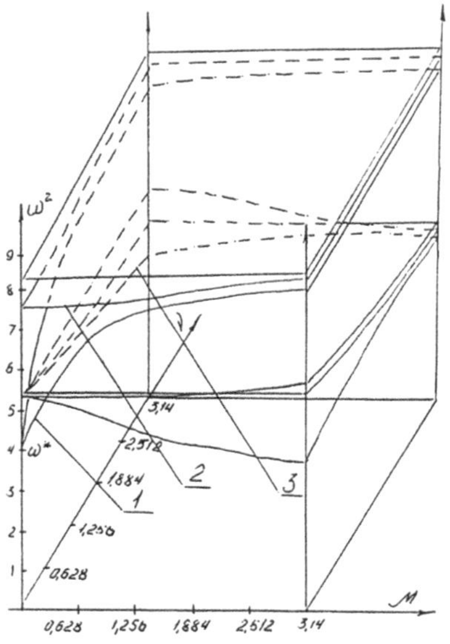

(a)

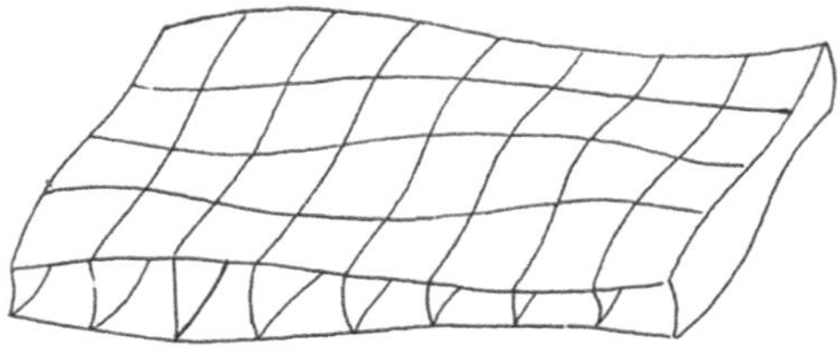

(b)

Fig. 3.12 a, b. Dispersion surfaces (**a**) and skew-symmetric modes (**b**)

bending modes is located below the other surfaces. This is why the bending mode has the lowest vibration frequency.

Figure 3.13 shows surfaces at imaginary values of μ and ν. The absence of bending vibrations with an imaginary value of μ in the range 0 to μ^* is their characteristic feature. The width of this area depends on the ratio between the bending stiffness of the outside laminas and the ribbings. The condition $\partial Z/\partial x = 0$ for the nodes at these modes is satisfied in this range because the stiffness of the ribbings is greater than the stiffness of the outside laminas. At point μ^* these stiffness are equal to each other.

It can be seen from the presented Figures that laminar ribbings have bending and shear deflections and their contribution to the stiffness of the system is different. This can be seen very clearly from the dispersion surface of the symmetric vibrations for the bending modes. This influence increases sharply with an increase in ν, i.e., the stiffness of the system along the ribbings is considerably higher compared to

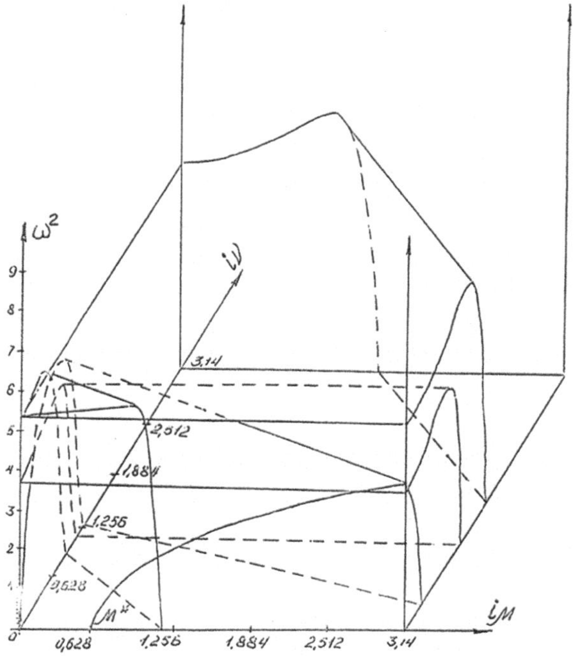

Fig. 3.13 Dispersion surfaces for imaginary values of μ and ν

the other directions. The stiffness of the system increases less along the x axis. This increase is determined by the bending stiffness of the ribbings and depends on the number of ribbings n comprised in a wavelength L. This influence is reduced with a decrease in n; at $n \rightarrow 0$, it tends toward zero. When calculating laminar systems with a crossing set of stiffness (laminar) ribbings, it is possible not to take into account the bending stiffness of the ribbings because the shifting stiffness is considerably higher than the bending one. The presence of stiffening ribs leads to the simultaneous existence, within some frequency range, of dispersion surfaces describing transversal motions along the z axis and a rotation about the x and y axes. Therefore, in this range, the bending modes, which are determined by the stiffness of the outside laminas, will get superimposed onto the bending modes determined by stiffness of external plates. Therefore, a local distortion of the basic mode will exist [1].

Let us now determine the natural frequencies of a system with specified boundary conditions using the dispersion equations. The solutions of our equations have the following form:

$$\mathbf{Z} = \mathbf{C}e^{i(\mu R+\nu s)+\omega t}$$

at both real and imaginary values of μ and ν; thus the sign at μ and ν can be anyone. Therefore, we will search for a solution in the form of their linear combination:

$$
\begin{aligned}
\mathbf{Z} &= \left[\mathbf{C}_1 e^{i\mu({}_1R+\nu_1 S)} + \mathbf{C}_2 e^{i(\mu_1 R-\nu_1 S)} + \ldots + \mathbf{C}_7 e^{i(-\mu_2 R+i\nu_2 S)} + \mathbf{C}_8 e^{i(-\mu_2 R+i\nu_2 S)}\right] e^{\omega t} = \\
&= \left[\left(\mathbf{A}_1 e^{i\nu_1 s} + \mathbf{A}_2 e^{-i\nu_1 s}\right)\left(\mathbf{B}_1 e^{i\mu_1 R} + \mathbf{B}_2 e^{-i\mu_1 R}\right)\right] e^{\omega t} \\
&+ \left[\left(\mathbf{A}_3 e^{\nu_2 S} + \mathbf{A}_4 e^{-\nu_2 S}\right)\left(\mathbf{B}_3 e^{\mu_2 R} + \mathbf{B}_4 e^{-\mu_2 R}\right)\right] e^{\omega t} = \\
&= \left(\mathbf{U}_R \mathbf{W}_S + \mathbf{U}'_R \mathbf{W}'_S\right) e^{\omega t}.
\end{aligned}
\tag{3.18}
$$

The coefficients B_3 and B_4 are equal to zero for the bending modes in the range of values $0 < \mu < \mu^*$ at $\nu = 0$. Indeed, in this range there are no solutions for this mode with imaginary values of μ_2. In the range $0 < \mu < \mu^*$, the boundary conditions of the rigid support type are implemented by laminar ribbings, and in the remaining area these ribbings implement an elastic support to the boundary nodes.

3.4.2 Vibrations of a Single-Section Lamina with Laminar Ribbing: Comparison of Discrete and Continuous Models

Let us determine the two lowest natural frequencies for a single-section lamina with a laminar ribbing. This system is shown in Fig. 3.14. It has the previous parameters. The boundary nodes are free.

Let us write the boundary conditions for the nodes $R = 0$ and $R = 8$ at $S = 0$:

$$
\left.\begin{aligned} Z_{00} &= 0 \\ Z_{08} &= 0 \end{aligned}\right\}
\begin{aligned} (\mathrm{Mom})_0 &= 0 \\ (\mathrm{Mom})_8 &= 0 \end{aligned}
$$

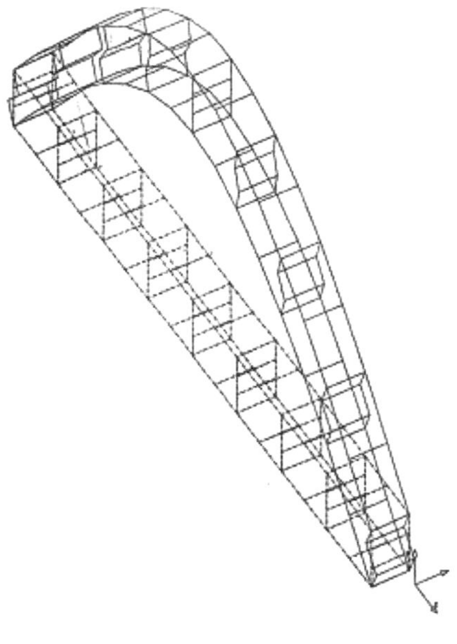

Fig. 3.14 Vibrations of a single-section lamina with a laminar ribbing

where $(\mathrm{Mom})_i$ is the bending moment at point *i*.

By substituting solution (3.18) into the boundary conditions, we obtain the frequency equation

$$\sin \mu R = 0 \quad (R = 8).$$

It follows from this that $\mu = \pi n/8, n = 1 \ldots \infty$. For the two lowest frequencies we obtain the values $\mu_1 = \pi/8$ and $\mu_2 = \pi/4$; to these correspond the natural frequencies of 0.209 and 0.75 Hz on the dispersion surface of the bending modes.

For the sake of comparison, the same frequencies were calculated using the FEM. Figure 3.14 presents the finite-element model and the normal modes for the lowest frequency. The system was decomposed into 108 nodes (648 degrees of freedom). The difference of the obtained values for the frequencies is insignificant (0.21 and 0.73 Hz); the first frequency deviates by 0.47% and the second one by 2.73%. However, in order to calculate these frequencies by means of the dispersion equations, it was sufficient to solve the natural value problem for a sixth-order matrix, while by using the FEM the same problem was solved for a matrix with a more than 100 times larger dimension.

Let us now study the replacement of such laminar systems with equivalent orthotropic lamina with a constant thickness. The closeness of the dispersion surfaces of the original system and the orthotropic lamina can serve as an equivalence criterion for such a replacement. The presence of skew-symmetric modes in the discrete model immediately sets a constraint on the possibility for continualization. Actually, these vibration modes, whose emergence is based on the ribbings vibrations, do not exist at all in continual lamina, which leads to a loss of this portion of the frequency spectrum. However, in the frequency spectrum where no skew-symmetric forms are present (from 0 to μ^*, Fig. 3.11), replacement is possible if the equivalence criterion is fulfilled. Therefore, let us compare the dispersion surface for the bending vibrations in these two systems. The vibration equation for a lamina with a constant thickness has the form

$$R_{11}\frac{\partial^4 w}{\partial x^4} + 2R_{12}\frac{\partial^4 w}{\partial y^2 \partial x^2} + R_{22}\frac{\partial^4 w}{\partial y^4} + \rho b\frac{\partial^2 w}{\partial t^2} = 0,$$

where b is the thickness of the lamina, ρ is the mass density of the material, R_{11} and R_{22} are the rigidities along the x and y axes, respectively, and R_{12} is the mixed stiffness.

The numerical dependence of R_{11}, R_{12}, and R_{22} on the value of $\rho\, b$ for some points of the dispersion surface of the bending modes of a system with ribbings for the above parameter values is shown in Table 3.1.

The values R_{11}, R_{12}, and R_{22} are not constant; they change noticeably from one point to another.

$$B = 100\,\mathrm{mm},\ b_{пл} = 5\,\mathrm{mm},\ b_{реб} = 4\,\mathrm{mm},\ L = 500\,\mathrm{mm}.$$

Table 3.1 Comparison of discrete and continuous models

№	Quantity rib in wave length	$\mu \times 10^{-1}$	ω^2			$R\rho b^{-1}$
			$\nu = 0$ $\omega^2 \times 10^{-3}$	$\mu = \nu$ $\omega^2 \times 10^{-5}$	$\mu = 0$ $\omega^2 \times 10^{-5}$	$R_{11} \times 10^{-3}$
1	101	0.314	47.97	0.7117	0.96	4934.6
2	54	0.628	67.7	3.073	2.389	4352.6
3	34	0.952	103.4	6.836	5.394	1313.1
4	26	1.26	155.2	11.9	9.632	623.63
5	21	1.7	220.5	18.0	13.13	362.92
6	17	1.884	299.8	25.02	21.89	237.97
7	15	2.198	395.5	32.62	23.64	169.4
8	13	2.12	508.0	40.89	24.64	127.8
9	12	2.826	681.5	49.91	28.58	99.01
10	11	3.14	775.5	59.81	35.24	79.78

In this way, the obtained results show that continualization is possible only in a very narrow frequency range. The width of this area and its location depend on the parameters of the system, the position of the ribbings, etc. The possibility for continualization of similar structures depends also on the ratio between the rigidities of the outside laminas and the internal ribbings. The internal ribbings have mainly shifting deformations when the lamina is being bending. Therefore, the deformation of the external and internal laminas is described by differential equations of different orders (fourth and second, respectively).

However, for a lamina supported by beam stiffness ribbings (Fig. 3.15), the deformation of the lamina and the ribbings has the same character. Therefore – as is known from [36] – continualization in the form of an orthotropic lamina is possible, which is confirmed by the type of dispersion surface (Fig. 3.15b).

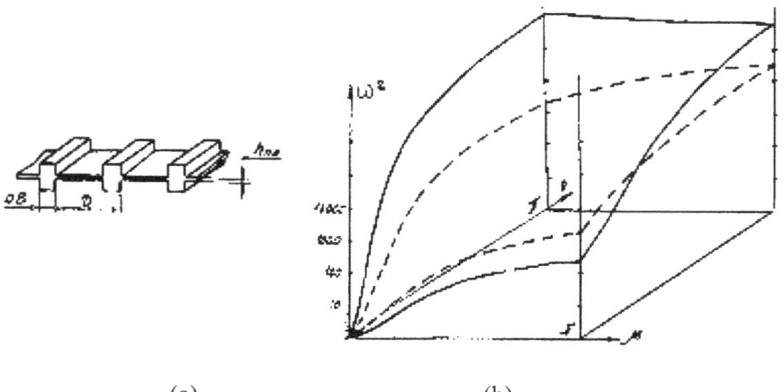

(a) (b)

Fig. 3.15a, b A plate with beam stiffness ribbings (**a**) and its dispersion surface (**b**)

3.5 Finite-Element Models for Beam Systems: Comparison with Distributed Parameters Models

3.5.1 Dispersion Equation for FE-Model of Beam

Let us consider a beam divided into $n+\ 1$ finite elements (Fig. 3.16).

The equation of the free vibrations of this system in the vertical plane, for s th node, has the following form:

$$\begin{aligned}&\mathbf{K}_{12}\mathbf{x}_{s-1}+(\mathbf{K}_{11}+\mathbf{K}_{22})\,\mathbf{x}_s+\mathbf{K}_{21}\mathbf{x}_{s+1}+\\&\mathbf{M}_{12}\ddot{\mathbf{x}}_{s-1}+(\mathbf{M}_{11}+\mathbf{M}_{22})\,\ddot{\mathbf{x}}_s+\mathbf{M}_{12}\ddot{\mathbf{x}}_{s+1}=0.\end{aligned} \tag{3.19}$$

where $\mathbf{x}_s$ is a two-dimensional vector with coordinates $[y_s, \theta_s]$, y_s is the vertical displacement of the node s and θ_s is the rotation angle. We search for a solution in the following form:

$$\mathbf{x}_s=\mathbf{C}\mathbf{z}^s e^{i\omega t}. \tag{3.20}$$

From Eq. (3.19):

$$\mathbf{D}_{12}\mathbf{z}^{-1}+(\mathbf{D}_{11}+\mathbf{D}_{22})+\mathbf{D}_{21}\mathbf{z}=0 \tag{3.21}$$

$$\mathbf{D}_{ij}=\mathbf{K}_{ij}-\omega^2\mathbf{M}_{ij},$$

where $\mathbf{z}$ is the root of the matrix equation (3.21).

In this way, we reduce the solution of the initial equation (3.19) of the order $2n$ to a solution of Eq. (3.21) of the order equal to 2.

Without taking the damping into account, solution (3.20) has the following form:

$$\mathbf{x}_s=\mathbf{C}_s e^{\pm\mu_r s}e^{i\omega t}=\mathbf{C}'_s e^{\pm\mu'_r l s+i\omega t} \tag{3.22}$$

where μ are real or purely imaginary values and $\mu' l=\mu$, l is finite element length. Solution (3.22) can be treated as a sum of a traveling and a reflected wave:

$$\mathbf{x}_s=\mathbf{C}_s e^{\pm\mu'_r l_s+i\omega t}.$$

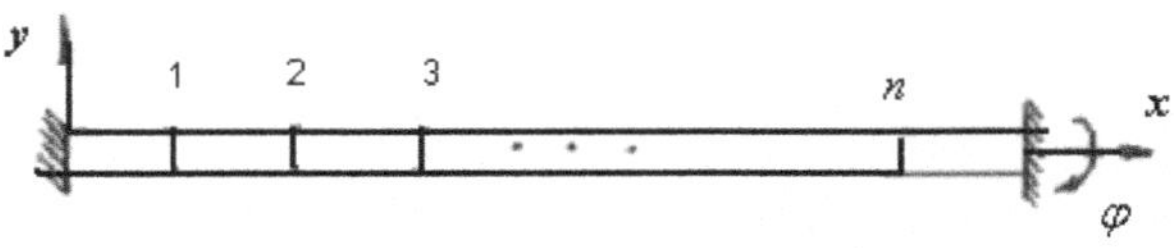

Fig. 3.16 A beam divided into n finite beam elements

For now, we will not impose any constraints on the boundary conditions of the system in Fig. 3.16. From Eq. (3.19) we obtain an equation for the node s. The expressions for the stiffness matrix $\mathbf{K}$ and the inertia matrix $\mathbf{M}$ for a beam finite element are provided in Chap. 2. Substituting (3.22) into (3.21) and performing certain transformations, we obtain the following dispersion equation:

$$\det\begin{vmatrix} \left(24-\alpha_0\omega^{*2}\right)-2\left(12+\alpha_1\omega^{*2}\right)ch\mu & 2\left(6+\alpha_2\omega^{*2}\right)sh\mu \\ -2\left(6+\alpha_2\omega^{*2}\right)sh\mu & \left(8-\alpha_4\omega^{*2}\right)+2\left(2+\alpha_3\omega^{*2}\right)ch\mu \end{vmatrix}=0, \tag{3.23}$$

where $\omega^{*2}=\omega^2\dfrac{ml^4}{EJ}$, $\alpha_0=\dfrac{26}{35}$, $\alpha_1=\dfrac{9}{70}$, $\alpha_2=\dfrac{13}{420}$, $\alpha_3=\dfrac{1}{140}$, $\alpha_4=\dfrac{2}{105}$.

Analogously to (3.10) wave parameter μ can be either real or imaginary

$$\mu^{1,2}=\pm\mu_1,\ \mu^{3,4}=\pm\mu_2,$$

by this for periodic solution μ_1 may be real and μ_2 be purely imaginary, i.e., $\mu_2=i\left|\mu_2\right|$. And $ch\mu_2=\cos\left|\mu_2\right|$, $sh\mu_2=i\sin\left|\mu_2\right|=\sin\mu_2$.

Dispersion equation (3.23) contains only the wave parameter μ_r and the dimensionless angular velocity ω^*. That is why it is correct for any parameters of beam EJ, l for any number of finite elements n, and, in addition, it does not depend on boundary conditions. The solution of this equation is presented in Table 3.2 and in Fig. 3.17.

If a solution in the form of Eq. (3.22) exists, it satisfies equation (3.23). Then the general solution of Eq. (3.19) can be presented in the form

$$\begin{aligned} x_s &= C_1e^{\mu_1 s}+C_2e^{-\mu_1 s}+C_3e^{i\mu_2 s}+C_4e^{-i\mu_2 s} \\ \theta_s &= \beta_1\left(C_1e^{\mu_1 s}-C_2e^{-\mu_1 s}\right)+\beta_2\left(C_3e^{i\mu_2 s}-C_4e^{-i\mu_2 s}\right), \end{aligned} \tag{3.24}$$

Table 3.2 FEM solution

ω^{*2}	s	μ_1	μ_2	β_1	$i\beta_2$	$\mu_1 n$	$\mu_2 n$
0.05	1	0.474	0.474	0.46	0.46	4.74	4.74
1.978	1*	1.185	1.185	1.185	1.184	1.186	4.74
4.015	4	1.413	1.413	1.413	1.413	14.13	14.13
15.3	2*	1.964	1.966	1.964	1.964	7.860	7.864
13.93	7	2.347	2.353	2.270	2.270	23.47	23.53
53.5	8	2.65	2.667	2.644	2.56	26.50	26.67
60.3	3*	2.715	2.743	2.708	2.57	10.86	10.97
82.79	9	2.917	2.962	2.88	2.34	29.17	29.62
216	4*	3.512	3.65	3.45	5.39	14.04	14.60
588.5	5*	3.987	4.49	3.94	7.93	15.95	17.96
990	15	4.125	4.955	4.118	10.926	41.25	49.55

Table 3.2 contains solution for n = 3 (marked*) and n = 9

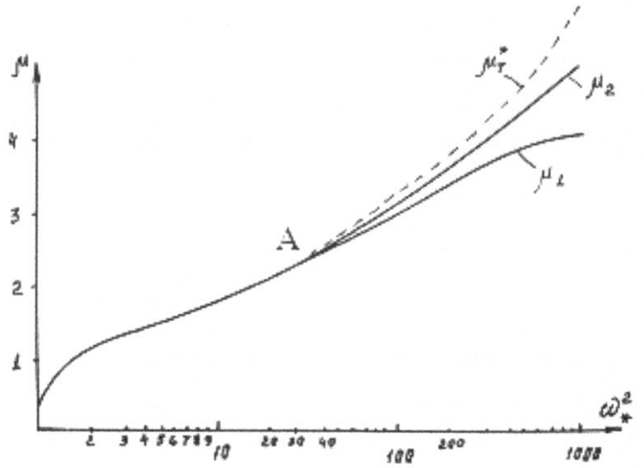

Fig. 3.17 Dispersion curve for finite-element model of a beam and a Bernoulli beam (*dashed line*)

Table 3.3 The solution for a Bernoulli beam with distributed parameters

$\overline{\mu}$	$\overline{\omega}^2$	$\Delta\omega$ %
4.73	500.55	
4.73	500.546	0.7
14.136	39943.8	0.4
7.853	3803.14	1.5
23.56	308106.7	1.8
26.70	505931.9	2.7
10.99	14587.83	2.9
29.841	792964	
14.136	39930.6	2
17.27	88954.8	
48.7	5624913	

where from Eq. (3.23)

$$\beta_r = -\frac{2\left(6 + x_2\omega^{*2}\right) sh\mu_s}{\left(24 - \alpha_0\omega^{*2}\right) - 2\left(12 + \alpha_1\omega^{*2}\right) ch\mu_s}, \ (r = 1, 2). \quad (3.25)$$

Thereby, it has been taken into account that $ch\mu_2 = \cos|\mu_2|$, $sh\mu_2 = i\sin|\mu_2| = \sin\mu_2$.

The constants C_j are determined, as is usually done, from the boundary conditions. The determinant Δc of this system gives the frequency equation for determining μ_r.

In particular, in the case of fixed ends, the boundary conditions $x_0 = \theta_0 = \theta_{n+1} = x_{n+1} = 0$ and the frequency equation will be as follows:

$$\Delta c = \left(\beta_1^2 - \beta_2^2\right) ch\mu_1 (n+1) \cos\mu_2 (n+1) = \\ 2i\beta_1\beta_2 sh\mu_1(n+1)\sin\mu_2(n+1). \quad (3.26)$$

The vibration modes are determined from Eq. (3.24) by taking into account the frequency determinant Δc and boundary conditions:

$$\begin{aligned} x_s = & -\beta_2 \cos \mu_2 (n+1)\, sh\mu_1 r - i\beta_1 \sin \mu_2 (n+1)\, r - \beta_1 sh\mu_1 (n+1-r) - \\ & - i\beta_2 sh\mu_1 (n+1) \sin \mu_2 r \\ \theta_s l = & \beta_1 [-\beta_2 \cos \mu_2 (n+1)\, ch\mu_1 r + i\beta_1 \sin \mu_2 (n+1) - \\ & - \beta_2 ch\mu_1 (n+1) \cos \mu_2 r + i\beta_1^2 sh\mu_1 (n+1) \sin \mu_2 r + \beta_1\beta_2 ch\mu_1 (n-1+r) + \\ & + \beta_1\beta_2 \cos \mu_2 (n+1-r)]. \end{aligned} \tag{3.27}$$

The solutions for other boundary conditions can be obtained analogously. Thus the frequency equation in the case of a hinge support is

$$\det \begin{vmatrix} sh\mu_1 (n+1) & i \sin |\mu_2 (n+1)| \\ -\left(6+\alpha_2\omega^{*2}\right) sh\mu_1 n + \beta_1 \left[\left(2+\alpha_3\omega^{*2}\right) \right. & -i\left(6+\alpha_2\omega^{*2}\right) \sin |\mu_2 n| + \beta_2 \left[\left(2+\alpha_3\omega^{*2}\right) \right. \\ \left. ch\mu_1 n + \left(4+\alpha_4\omega^{*2}\right) \times ch\mu_1 (n+1)\right] & \left. \cos \mu_2 n + \left(4+\alpha_2\omega^{*2}\right) \cos \mu_2 (n+1)\right] \end{vmatrix} = 0. \tag{3.28}$$

3.5.2 *Comparing Models with Distributed Parameters and Finite Elements Models at Different FE-Mesh*

Let us now compare the solutions obtained for a beam with distributed parameters and one based on finite elements. The equation for a Bernoulli beam with distributed parameters is

$$\mathrm{E}J\overline{x}^{IV} + m\ddot{\overline{x}} = 0. \tag{3.29}$$

[the index "–" corresponds here to solution (3.29)].

The natural frequency $\overline{\omega}$ and the wave parameter $\overline{\mu}$ for Eq. (3.29) are related by the known relation [20,106]

$$\overline{\mu}^4 = \frac{m\overline{\omega}^2}{\mathrm{E}J} \tag{3.30}$$

This relation does not depend on the boundary conditions. Let us find now the ratios between $\overline{\omega}$, $\overline{\mu}$, and the quantities ω^*, μ_r $(r = 1{,}2)$ included in Eq. (3.23).

Equation (3.30) determines four roots: $\overline{\mu}^{1,2} = \pm|\mu|$, $\overline{\mu}^{3,4} = \pm i|\mu|$. In the sequel, $\overline{\mu}$ will be understood to mean the root $\overline{\mu} = |\overline{\mu}|$. To begin with, let us compare the dispersion equations (3.29) and (3.30). It should be noted that FEM solution (3.23) coincides with solution (3.30) if we assume that $|\mu_r|\, s = \bar{\mu} x_s$, i.e., if $\mu_r (n+1)$ corresponds to $\overline{\mu}L$. Therefore, in order to do the comparison, we will assume

$$\bar{\mu} l = \mu_r \text{ or } \bar{\mu} z = \mu_r (n+1)\,;\; L = l(n+1). \tag{3.31}$$

Taking into account (3.31), as well as the expressions for ω^*, we note that to do the comparison with (3.25) it is necessary to rewrite Eq. (3.30) in a coordinates:

$$\overline{\mu}^* = \overline{\mu} l; \ \overline{\omega}^{*2} = \overline{\omega}^2 \frac{ml^4}{EJ}, \tag{3.31a}$$

Multiplying the two sides of (3.31) by l^4, we obtain

$$\overline{\mu}^{*4} = \overline{\omega}^{*2}, \tag{3.32}$$

which is obviously correct at any l, and, therefore, at any n. Curve (3.32) is shown in Fig. 3.17 with a dotted line.

The curves in Fig. 3.17 do not depend on the parameters of beam EJ, l, or on the number of finite elements n, or on the boundary conditions.

Let us find now the area where the FEM dispersion equation (3.23) and the equation for a beam with distributed parameters (3.32) coincide. In accordance with Vieta's theorem, we obtain from Eq. (3.23)

$$ch\mu_1 + \cos\mu_2 = \frac{\left(24 - \alpha_0\omega^{*2}\right)\left(4 + 2\alpha_3\omega^{*2}\right) - \left(12 - 2\alpha_1\omega^{*2}\right)\left(8 - 2\alpha_4\omega^{*2}\right)}{4\left(6 + \alpha_2\omega^{*2}\right)^2 - \left(24 - 2\alpha_1\omega^{*2}\right)\left(4 + 2\alpha_3\omega^{*2}\right)}.$$

Assuming $\mu_1 = |\mu_2|$ and neglecting the terms of the order of $\mu^8/8$ in the expansion in a series in powers of μ (at $|\mu| = 2.2$ this gives an error of $0.02 << 1$), we obtain

$$\mu^4 = \omega^{*2} \frac{1 + \omega^{*2}/735}{1 + \dfrac{\omega^{*2}}{420} + \dfrac{7\omega^{*2}}{12\,(420)^2}}.$$

which at $\omega^* = 15$ (point A in Fig. 3.17) coincides with the solution of Eq. (3.32) for continuous beam with an accuracy of up to 1.5%.

Until the final inference regarding the comparison of the solutions, it is still necessary to evaluate the level of closeness of the vibration modes and the frequency equations.

The frequency equation for FEM (3.26) coincides with the frequency equation for a beam with distributed parameters in the case of fixed ends

$$1 - \cos\overline{\mu}z ch\overline{\mu}z = 0,$$

if

$$\beta_1 = i\beta_2 = \mu_1 = i\,|\mu_2| = \overline{\mu}. \tag{3.33}$$

It can be shown that the FEM frequency equation (3.28) for a beam with hinge-supported ends also coincides with the known equation for a beam with distributed parameters

$$\sin \mu_2 (n+1) = \sin \mu_n z = 0$$

under the same conditions (3.33).

Analogous conclusions will be correct also in the case of other boundary conditions. Under condition (3.33), the expressions for the vibration modes of FEM-model and with distributed parameters one (3.24) and (3.27) also coincide with those known for a beam with distributed parameters.

Therefore, conditions (3.31), as well as (3.33), which follow from them, determine the area of coincidence of these models. A portion of the dispersion curve from the coordinate origin up to point A satisfies these conditions, where coordinates of point A:

$$A\left(\mu_A = 2.2;\ \omega_a^{*2} = 15\right), \text{ (Fig. 3.17, Tables 3.2, 3.3).}$$

If $|\mu| > 2.2$, then $|\mu_1| \neq |\mu_2|$, although this difference is insignificant up to $\mu \cong 2.7$.

Therefore, the comparison of the solution for the finite-element model of a beam with that for a beam with distributed parameters permits one to make some recommendations regarding the necessary quantity of finite elements:

1. At sufficiently small values of (μ, ω^{*2}), dispersion curve (3.23) for FEM model coincides with Eq. (3.32) for a distributed model. This coincidence is sufficiently close at $\mu < 2.2$; in this area, $\mu_1 = |\mu_2|$ and the frequency equations also coincide. With the increase of μ, the difference between these solutions can become significant.
2. Determine the areas where the two solutions do not coincide.
   At $98.8 < \omega^{*2} < 120$, there is a filter zone (no-pass band) for the FEM model of beam. In this zone, both roots μ_0 and μ_2 are real and there is no harmonic component. This fact reflects the wave nature of the FEM model. However, such a zone cannot exist in a system with distributed parameters, which is confirmed also by Tables 3.2, 3.3.
3. Determine the necessary number of finite elements n as a function of the admissible error $\Delta\omega$ % in the calculation of the natural values in the range 0 to $\omega_{\max}$.
   To this end, it is necessary to find the dimensionless quantity $\omega^{*2} = \dfrac{\omega_2 z^4}{n^4 mEJ}$, which depends on two parameters, ω^2 and n.
   This makes it possible to solve the following two problems:
   - At a given maximum frequency $\omega_{\max}$, determine the necessary number of finite elements n that guarantee a solution coincidence accuracy of $\Delta\omega\%$;
   - On the given number of finite elements n, determine the frequency limit $\omega_{\max}$ within which the coincidence accuracy of $\Delta\omega\%$ for the necessary solutions is guaranteed.

It should be pointed out that the FEM frequency equation (3.26) gives values of the wave parameter μ that are sufficiently close to $\overline{\mu}$. Indeed, Table 3.2

contains the values $\mu_1 (n+I)$ and $\mu_2 (n+I)$ for $n = 3$ (marked with $*$) and $n = 9$; p is the number of mode and $\overline{\mu}z$ the known solution for a beam with distributed parameters at fixed ends. From Tables 3.2, 3.3

$$(\overline{\mu}z - \mu_1 (n+1)) / \overline{\mu}z \leq 1\%$$

and

$$(\overline{\mu}z - \mu_2 (n+1)) / \overline{\mu}z \leq 0.1\% \text{ at } \omega^{*2} < 60.$$

4. Find the relationship between the number of finite elements n and the necessary maximum number of oscillation forms $p_{\max}$.

To this end, it is necessary to plot a graph $(\Delta\omega\%, \mu_r)$ based on Tables 3.2, 3.3.

As has already been pointed out, the values μ_1, μ_2 at $\mu < 2.2$ coincide well with the values for a beam with distributed parameters; as is known [20] for a beam with fixed ends, the wave parameter for the p th mode is

$$\overline{\mu}_p L = \frac{2(p+1)+1}{2}\pi (p > 2), \tag{3.34}$$

$$\bar{\mu}_1 L = 4.73.$$

From Fig. 3.17 and on the basis of a given coincidence accuracy $\Delta\omega\,\%$, we determine the corresponding value $\mu_\Delta (\Delta\omega\%)$. Taking into account Eq. (3.31), which is fulfilled at $\mu_{1,2} < 2$, we obtain from Eqs. (3.31) and (3.34)

-for a beam with fixed ends

$$n_{\text{fix}} = \frac{(\mu L)_{\max}}{\mu_\Delta} = \pi \frac{2(p+1)+1}{2\mu_\Delta}, \tag{3.35}$$

– for a hinge-supported one

$$n_h = \frac{p+1}{\mu_\Delta}\pi, \tag{3.35a}$$

where p is the number of modes (which coincides with the number of nodes in the mode of interest to us), μ_Δ is the value μ that corresponds to the given value of the accuracy $\Delta\omega\,\%$.

For example, at $p = 5$, $n_{ш} \cong n_{защ} = 7$:at n = 10, $n_{ш} \cong n_{защ} = 14$.

In this way, the recommendation regarding the selection of the number of finite elements of the beam as a function of the given coincidence accuracy is defined by ratio (3.35). It should be pointed out that Fig. 3.17 shows, for simplicity's sake, the approximate ratio (3.35), where it is assumed that $\mu_\Delta \cong \overline{\mu}$, which – as was shown above – is fulfilled with a sufficient precision at $\mu < 2.2$.

Analyzing (3.35), we will mention the following important fact: in practical terms, the boundary conditions do not affect the number of finite elements n. Indeed, from Eq. (3.35): independently of the number of forms p.

$$n_{защ} - n_{ш} = \frac{\pi}{2\mu_{\Delta}} = 1.$$

For instance, at $p = 5$ and $\Delta\omega = 3\%$, we obtain from Eq. (3.35) and (3.35a):

$$n_{защ} = n_{ш} + 1 = 7; \text{ at } p = 10 : n_{защ} = n_{ш} + 1 = 13.$$

It is according to principle Sent–Vienan about small area of influence of conditions of fastening

When it is necessary to take into consideration the high-frequency vibrations for obtaining the respective evaluations, it is necessary to consider the Timoshenko–Rayleigh beam [106] instead of Bernoulli beam (3.29).

3.5.3 Beam Systems

It is very important for applications to have available a generalization of these results for systems consisting of several beams (such as, for example, in Fig. 3.18).

The first problem for a system consisting of q beams (Fig. 3.18) is the selection of an adequate number of finite elements for each beam. This means the number of finite elements where each beam ensures the same precision of numerical calculation in the given frequency range. In fact, even if only one j th beam gives a numerical error $\delta_j > \delta_0$ (where δ_0 is the required calculation accuracy for the whole system), then the numerica error for the whole system will be $\delta = \delta_j$. On the other hand, the unnecessarily large number of finite elements will not improve

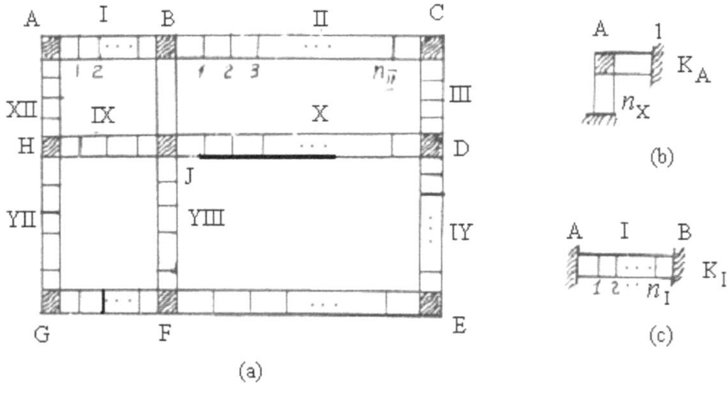

Fig. 3.18a–c A beam system (**a**) and its subsystems kinds (**b**), (**c**)

the numerical accuracy of the system and will lead only to an unreasonably high volume of numerical work.

The research that was performed also makes it possible to give an answer to this question. It can be shown that the number of finite elements for each beam of a system will be coordinated within the given calculation accuracy δ_0 if at $\omega < \omega_{\max}$ (where $\omega_{\max}$ is the upper limit of the frequency range of interest to us) the following condition is fulfilled for each beam:

$$\mu_{j\max} \leq \mu_{j0} \text{ at } \omega^{*2}_{\max} = \frac{\omega^2_{\max}}{EJ_j/l_j^4} \quad (j = 1, \ldots, q), \tag{3.36}$$

where μ_{j0} is the value μ that ensures the required accuracy of the calculation (a value μ_{j0} is determined for each beam from Table 3.2). From Eq. (3.36) it follows that the number of finite elements n_j for each beam is such that

$$\frac{\omega_{\max} l_j^2}{\sqrt{EJ_j}} = \frac{\omega_{\max} l_j^2}{\sqrt{EJ_j} n_j^2} = \omega^{*2}_{\max}. \tag{3.37}$$

In this way

$$n_1^2 : n_2^2 \cdots = \frac{l_1}{\sqrt{EJ_1}} : \frac{l_2}{\sqrt{EJ_2}} = \cdots,$$

where l_j is the length of the jth beam.

In accordance with (3.37), the nearest whole number n_j is selected. Then $\omega^{*2}_{j\max}$ and, therefore, also $\mu_{j\max}$ are practically equal for each beam, which ensures an identical accuracy at the calculation at $\omega < \omega_{\max}$. The ratio (3.37) does not depend on the boundary conditions since it has been obtained on the basis of the dispersion equation. But the number of finite elements for the beam will depend on the boundary conditions. Here, just as above, the beam can be considered fixed and in this way the upper limit n_{fix}, can be obtained.

Condition (3.37) means that the partial frequencies, i.e., the frequencies of the nodes located on the main diagonal of stiffness matrix, are the same. It in turn is one of the criterias for the good conditioning of matrix **D** (Chap. 2, Sect. 2.1.8).

Condition (3.37) also makes it possible to identify the beams with a great discrepancy of elastic-inertial characteristics and to sort out those beams whose influence on the dynamics of the system in the given frequency range is slight. It is obvious that it will be fulfilled for such beams that cannot satisfy (3.37) even at $n = 1$. For matrix **D**, this would mean the existence of small energy and spectral coefficients of interaction (2.7)

$$\boldsymbol{\alpha}_{ij} < \varepsilon; \; \mathbf{s}_{ij} < \varepsilon \; (j \neq i).$$

This condition means that this beam has a weak dynamic interaction with the remaining part of the system. A correction to the natural frequencies and vibration modes was calculated in Chap. 2 and it was shown that it is possible not to take this correction into consideration in the first approximation (in the absence of multiple frequencies in partial subsystems).

3.6 Hierarchy of Mathematical Models: Superposition of Wave Motions

One of the most widely used approaches to the dynamic analysis of a big structure is the creating of a models hierarchy as a gradual refinement of the model details. Such an approach is convenient because it permits one to have available at the very beginning a sufficiently simple, although as yet approximated, picture of the phenomena. In the course of its gradual sophistication, it is possible to reveal the importance of those and other elements of the system. The frequency hierarchy of the models is very important. Let us now consider some rather simple examples of such model hierarchy illustrated in Figs. 3.19 and 3.20. These systems represent periodic structures in which further the elasticity of the rigid bodies is taken into account by representing them as masses linked with elastic elements. Such a refinement of the model is necessary for the study of the higher frequency range. Let us call the model in Fig. 3.19a the first level of the hierarchy and the one in Fig. 3.20a – the second level. These simple systems permit one, nonetheless, to establish general patterns to which similar refinements lead in the general case.

Let us prove the following affirmation: The higher-level models in a hierarchy describe the resulting vibrations in the form of superposition of wave motions corresponding to the preceding levels of the hierarchy. We will prove this first for regular systems of the type shown in Figs. 3.19 and 3.20a, b. The free vibrations for regular system in Fig. 3.19a (hierarchy of the first level) are described by equations of the following kind:

$$d_{s,s-1}x_{s-1} + d_{ss}x_s + d_{s,s+1}x_{s+1} = 0, \tag{3.38}$$

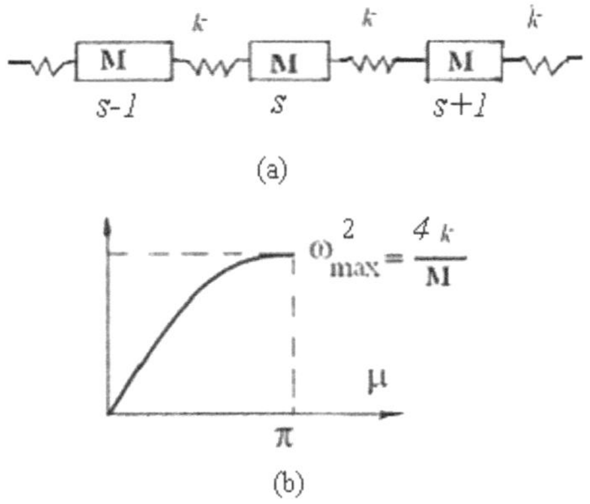

Fig. 3.19a, b Model of the first level in the hierarchy (**a**) and its dispersion curve (**b**)

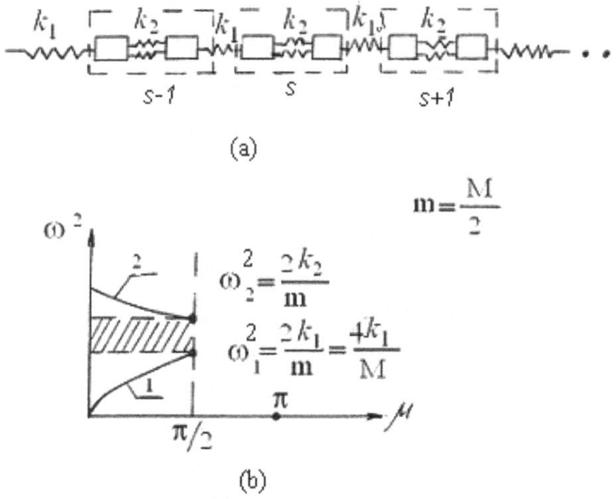

Fig. 3.20a, b Model of the second level in the hierarchy (**a**) and its dispersion curve (**b**)

where

$$d_{s,s+1} = d_{s,s-1} = -k;\ d_{ss} = 2k - M\lambda^2,\ (s = 1 \ldots n).$$

The solution for x_s has a wave character

$$x_s = C_1 e^{\mu s} + C_2 e^{-\mu s}. \tag{3.39}$$

The constants C_1, C_2 are determined from the boundary conditions at $s = 0$, $s = n + 1$. The dispersion equation is

$$ch\mu = 2 - \nu^2\lambda^2;\ \nu^2 = 2k/m.$$

The hierarchy of the second level takes into account the elasticity of each of the masses M, which is represented as a chain of t masses $m = M/t$ connected by elastic elements (springs) (Fig. 3.20); the total mass of the system does not change. In this case, all elements in Eq. (3.38) become blocks of the order of t

$$\begin{bmatrix} 0 & \cdots & -k_1 \\ 0 & \cdots & 0 \\ 0 & \cdots & 0 \end{bmatrix} x_{s-1} + \begin{bmatrix} -m\lambda^2 + k_1 + k_2 & -k_2 & 0 \ \cdots & \\ -k_2 & -m\lambda^2 + 2k_2 & -k_2 \ \cdots & \\ \cdots & \cdots & \cdots & \cdots \end{bmatrix} x_s +$$

$$+ \begin{bmatrix} 0 & \cdots & 0 \\ 0 & \cdots & 0 \\ 0 & \cdots & -k_1 \end{bmatrix} x_{s+1} = 0.$$

In addition, the coordinate x_s now becomes as a vector

$$\mathbf{x}_s = \left[y_{1s}, \ldots, y_{ts}\right], \ (s = 1 \ldots n),$$

where y_{rs} is the coordinate of the mass r in the s th integration. It is easy to see that the solution for y_{rs} has a wave character, and it can be sought in the form

$$y_{rs} = c_s^{(1)} e^{\mu_2 r} + c_s^{(2)} e^{-\mu_2 r}, \ (r = 1 \ldots t), \ (s = 1 \ldots n).$$

By accepting the boundary conditions

$$y_{os} = x_s, \ y_{t+1,s} = x_{s+1}.$$

we ensure the compatibility of the displacements. Then

$$c_s^{(1)} + c_s^{(2)} = c_1 e^{\mu_1 s} + c_2 e^{-\mu_1 s}$$
$$c_s^{(1)} e^{\mu_2 (t+1)} + c_s^{(2)} e^{-\mu_2 (t+1)} = c_1 e^{\mu_1 (s+1)} + c_2 e^{-\mu_1 (s+1)}.$$

These equations are satisfied at

$$c_s^{(1)} = c_1 e^{\mu_1 s}; \quad c_s^{(2)} = c_2 e^{-\mu_1 s} \quad \mu_2 = \frac{\mu_1 + 2\pi r}{t+1}, \quad (r = 1 \ldots t),$$

i.e.

$$y_{rs} = c_1 e^{\mu_1 s} e^{\mu_2 r} + c_2 e^{-\mu_1 s} e^{-\mu_2 r}.$$

Therefore, y_{rs} is represented as a superposition of two waves: with wave numbers μ_1 and μ_2. After taking into consideration that for the chain with t - mass there can be t values for μ_2, the solution for y_{rs} will have t branches for each of the s modes. And the displacements of the lower level model as some averaging of the higher level one may be considered.

For simplicity's sake, let us consider the system for $t = 2$ (Fig. 3.20). The dispersion equation for the first hierarchy level, Fig. 3.19, is

$$\lambda_2 = \frac{4\kappa}{M} \sin^2 \mu_1/2$$

and the corresponding vibration modes represent the harmonics.

For the second level of the hierarchy, the dispersion equation is

$$\begin{bmatrix} k_1 + k_2 - \frac{M}{2}\lambda^2 & -k_1(1 + e^{-\mu}) \\ -k_1(1 + e^{\mu}) & k_1 + k_2 - \frac{M}{2}\lambda^2 \end{bmatrix} = 0.$$

It is shown in Fig. 3.20b. Comparing Figs. 3.19b and 3.20b, we become convinced that at $\lambda \leq \lambda_1$ the dispersion equations coincide completely, and at $\lambda > \lambda_1$

the model of the first level does not work. At $\lambda_2 < \lambda < \lambda_1$, the model of the second level has a no-pass band and there are two independent branches of the dispersion equation: At the lower one, the two neighboring masses vibrate in phase, and at the upper one in antiphase; the resulting vibration mode is presented as a superposition of two waves with wave values equal to μ and π.

In this way, the model of the first level is acceptable up to a frequency $\omega_{max} = 4\ k_1/M$, which is equal to the doubled partial frequency of the mass M. For the section with a length L, the corresponding length of the longitudinal wave l is equal to $l = L/4$; therefore, the elasticity of the element may not be taken into account if its length does not exceed one fourth of the wavelength.

These pretty simple recommendations can be used also in the calculation of more complex systems with a hierarchical structure.

3.7 Vibrations of Self-Similar Structures in Mechanics

3.7.1 Self-Similar Structures: Basic Concepts

Regular structures with translational symmetry represent nonetheless a relatively limited class of structures and do not describe many natural and technical systems. However, this class can be expanded considerably by complementing it with a class of systems called self-similar structures (fractals). The fractal (from Latin *fractus*) is a structure consisting of parts with each one similar to the structure as a whole [32, 74, 75]. This is one of the most important principles of morphogenesis in nature; it is manifested in many objects of nature and in innumerable phenomena in the world around us (for example, the crowns of trees, the system of blood circulation, etc.).

There is a simple recursive procedure for obtaining fractal curves in a plane. Let us define an arbitrary broken line with a finite number of elements (this is the generator). Let us then replace in it each element with a generator in appropriate scale. By continuing this process, we will obtain a fractal curve (Koch's curve).

In particular, the following structures belong to self-similar structures:

- Structures in which each cell repeats on a certain scale the structure of the preceding one. The translational symmetry is accompanied here by a similarity transformation (scaling) of the neighboring cells. This is one of the fractal types;
- Structures with a spiral symmetry and positioning of the cells along a screw line spiral with a constant step. In particular, systems with a reflection symmetry belong to this structure;
- Self-similar structures in which the spiral symmetry is accompanied by a similarity transformation of the neighboring cells.

Such structures are widespread in biology, polymer chemistry, physics. The growth and formation of polymer molecules and crystals is accompanied by the formation of fractals. The theory of fractals is currently very popular; many scientists from various fields study it mainly as a method of forming various structures.

As regards the study of such structures in mechanical constructions, human-created structures are simpler than those of naturally occurring fractals. However, they have specific features and their dynamic and vibration properties are of paramount importance. The dynamic properties of self-similar structures are studied considerably less, and this section is devoted to them.

Among the self-similar structures widely used in mechanical engineering are, for example, bars with a stepwise change in their cross section, crankshafts (spiral symmetry), conical springs, etc.

Here we will study the main dynamic properties of such structures and the methods for studying sufficiently simple systems.

3.7.2 Vibrations in Self-Similar Mechanical Structures: Dispersion Equation

The mathematical study of the dynamic processes in self-similar structures can be done on the basis of the unified approach we are about to describe. Let us analyze the longitudinal vibration of a rod with a circular cross section and with masses fixed on it. We will assume that the geometric parameters of the rod (length l, radius of the cross section r) change within the same scale γ, with the value of the masses changing as well. Then the stiffness of cell $i+1$ is

$$k_{i+1} = \frac{EF_{i+1}}{l_{i+1}} = \frac{E\pi\gamma^2 r_i^2}{\gamma l_i} = \gamma k_i,$$

and for the masses $m_i = \gamma m_{i+1}$. In this way, the elastic and inertial characteristics of the system change within the same scale (Fig. 3.21). Such a structure can be considered as a type of fractal (single-scale).

The equation of such a self-similar structure in matrix form is

$$\begin{bmatrix} -m_1\lambda^2 + k_1 & -k_1 & & \\ -k_1 & -m_2\lambda^2 + k_1(1+\gamma) & -\gamma k_1 & \\ & -\gamma k_1 & -m_3\lambda^2 + k_1\gamma(1+\gamma) & -\gamma^2 k \\ & & -\gamma^2 k & \vdots \end{bmatrix} \mathbf{X} = 0, \tag{3.40}$$

where $\mathbf{X}^T = [x_1, x_2, x_3 \ldots x_n]$

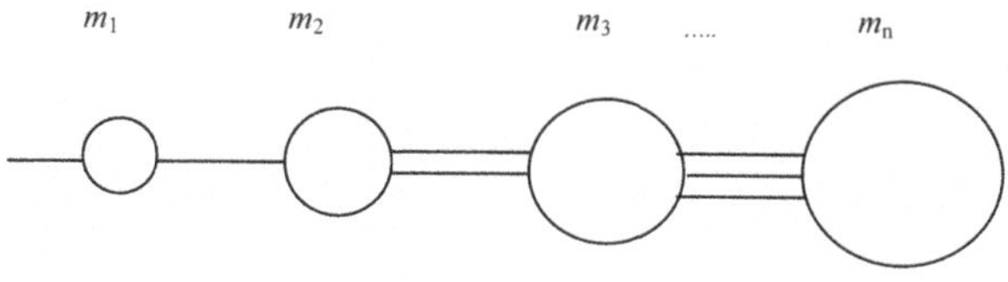

Fig. 3.21 Self-similar rod structure with masses

Let us now perform a substitution of the variables:

$$\mathbf{X}^* = \mathbf{N}\mathbf{X}, \quad \mathbf{N} = \begin{bmatrix} \frac{1}{\gamma} & & & \\ & \frac{1}{\gamma^2} & & \\ & & \ddots & \\ & & & \frac{1}{\gamma^n} \end{bmatrix}. \tag{3.41}$$

This transformation of the coordinates means multiplication of the matrix $[\mathbf{K} - \lambda\mathbf{M}]$ on the right and left by matrix $\mathbf{N}$. As a result we obtain

$$\left(\mathbf{K} - \lambda^2\mathbf{M}\right)^* = \mathbf{N}^T\left(\mathbf{K} - \lambda^2\mathbf{M}\right)\mathbf{N},$$

$$\left(\mathbf{K} - \lambda^2\mathbf{M}\right)^* = \begin{bmatrix} -m_1\lambda^2 + \frac{k_1}{\sqrt{\gamma}} & -\frac{k_1}{\sqrt{\gamma}} & & \\ -\frac{k_1}{\sqrt{\gamma}} & -m_1\lambda^2 + \frac{k_1(1+\gamma)}{\gamma} & -\frac{k_1}{\sqrt{\gamma}} & \\ & -\frac{k_1}{\sqrt{\gamma}} & -m_1\lambda^2 + \frac{k_1(1+\gamma)}{\gamma} & \\ & & & \vdots \end{bmatrix}$$

This equation describes a structure with translational symmetry (Fig. 3.22) with identical masses m_1 with the stiffness between them equal to $\frac{k_1}{\sqrt{\gamma}}$. And, in this case there is additional fixing of the masses equal to

$$\frac{k_1(1+\gamma)}{\gamma} - 2\frac{k_1}{\sqrt{\gamma}} = \frac{k_1(1-\sqrt{\gamma})^2}{\gamma} = k^*.$$

Therefore, all the above-described methods for studying periodic structures can be applied also to self-similar structures (after taking into consideration the transformation of coordinates (3.41)).

Since the linear transformation of coordinates (3.41) does not change the frequency properties, the structures in Figs. 3.21 and 3.22 have identical frequencies.

The dispersion equation for the system in Fig. 3.22 has the form:

–for real μ:

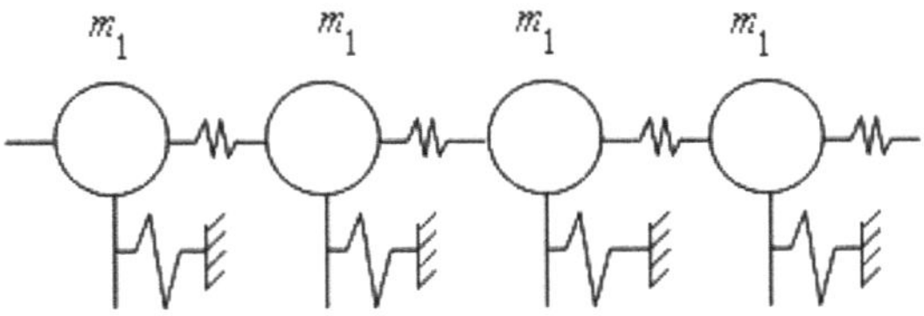

Fig. 3.22 Equivalent periodic structure with the same natural frequencies

$$-m_1\omega^2 + \frac{k_1(1+\gamma)}{\gamma} - 2\frac{k_1}{\sqrt{\gamma}}\cos\mu = 0, \tag{3.42a}$$

for purely imaginary $\mu = i\mu'$:

$$-m_1\omega^2 + \frac{k_1(1+\gamma)}{\gamma} - 2\frac{k_1}{\sqrt{\gamma}} ch\mu' = 0. \tag{3.42b}$$

It can be seen from these equations that the given system represents a mechanical frequency band filter. The bandpass of the harmonic signal is $\omega_0 < \omega < \omega^*$ (Fig. 3.23), where

$$\omega_0^2 = \frac{k^*}{m} = \frac{k\left(1-\sqrt{\gamma}\right)^2}{m\gamma} \text{ at } \mu = 0,$$

$$\omega^{*2} = \frac{k\left(1+\sqrt{\gamma}\right)^2}{m\gamma} \text{ at } \mu = \pi.$$

Therefore, the band pass is defined as

$$\Delta\omega^2 = \omega^{*2} - \omega_0^2 = \frac{4k}{\sqrt{\gamma}}.$$

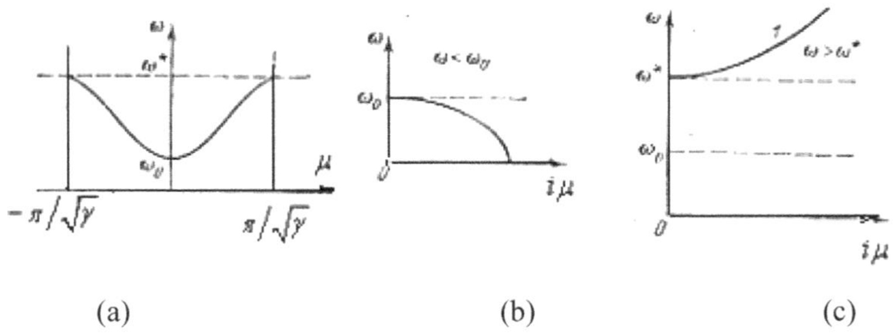

Fig. 3.23a–c Dispersion curves for real (**a**) and imaginary wave values (**b**), (**c**)

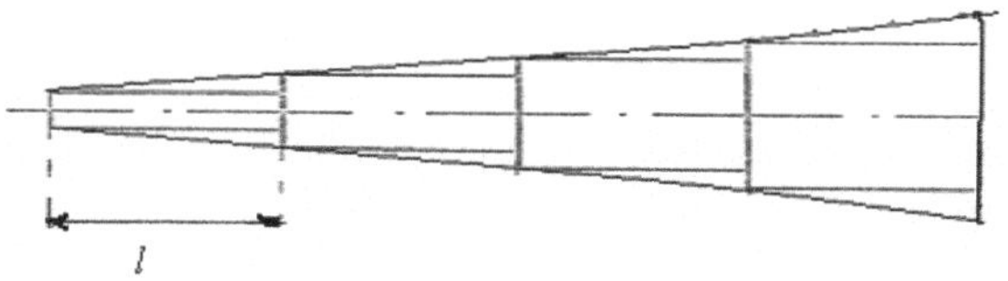

Fig. 3.24 A rod with a stepwise cross section

We can see that the band pass is inversely related to the similarity parameter γ. The larger γ, the lower the band passes.

This parameter is not very large for mechanical systems: on the order of 1.5–3. However, this parameter can be larger for biological systems. If γ is very large, then at the limit the band pass becomes very narrow, meaning that the system is set to a single frequency.

Let us consider some self-similar mechanical systems.

1. A beam system (Fig. 3.24) in which the parameters of the sth section change following an exponential pattern

$$T(s) = a^s = e^{bs}.$$

At the same time, the coefficients of scaling a (or b) can be different for the different parameters of the system:

$$r_s = r_0 a_1^s, \quad m_s = m_o a^s\,, \; a = a_1^2,$$

where r_s is the radius of the cross section s and m_s – is the mass of this section, which we consider to be concentrated in the center of the mass.

Then, obviously, $m_s = m_0 l\pi r^2$, and $m_{\,0}$ is the mass per unit length.

For longitudinal vibrations along the horizontal axis, we find that the stiffness of each section is

$$k_s = \frac{EF}{l} = \frac{E\pi a_1^{2s}}{l}.$$

If we accept that

$$a = a_1^2 \tag{3.43}$$

then the system becomes a single-scale one, as was discussed earlier. In this case, the partial frequencies $\nu_s = \dfrac{k_s}{m_s}$ are the same for each section, i.e.,

$$\nu_s = \frac{k_s}{m_s} = \frac{E\pi a_1^{2s}}{l m_0 a^s} = \text{const}.$$

By using a transformation of the coordinates according to condition (3.43),

$$x_s = a_1^3 x_s^*,$$

we obtain equations for the longitudinal vibrations in finite difference which are analogous to Eq. (3.42):

$$k_{s-1}x_{s-1} + (k_s - m_s\lambda^2)x_s + k_{s+1}x_{s+1} = 0 \quad (s = 1 \ldots n).$$

When we consider a beam in a limit at $n \to \infty$ (n is the number of sections), we obtain a beam with equal resistance. If its cross section changes along its length following an exponential pattern, then

$$F(s) = F \exp(\rho F_0(s)/P).$$

As is known, such a beam has the same static load in each cross section at a acting the constant load P (ρ is the density of the material).

In particular, the well-known "golden section," which has been used in architecture since ancient times, also follows an exponential law of change. This is a special case of a self-similar change in the parameters of the section according to the law

$$T(s) = a^s, \; a = \Phi = \sqrt{5} - 1.$$

The some other numerical results of this class structures, will be given below (Chap. 9) with help of dynamic compliances method.

Chapter 4
Vibrations of Systems with Geometric Symmetry. Quasi-symmetrical Systems

4.1 Introduction

Systems with symmetries of various types have long attracted the attention of scientists in all fields: chemistry, physics, and biology. Systems with geometric symmetry find wide application in many areas of engineering. They form various kinds of machine platforms, cyclically repeated forms of stators, rotors with blades mounted on them, etc. Symmetric systems are widely applied also in the civil engineering where their use is convenient due to technological process, type-design, and architectural requirements. There exists a large body of literature devoted to the study of the statics and dynamics of symmetrical structures. As a matter of fact, it is possible to identify a number of characteristic features in the dynamic behavior of symmetric systems that make their use particularly convenient. Thus, in particular, the following basic properties are typical for such systems:

– the presence of a "quiet" point – center of symmetry – insofar as this point is a node in rotational vibrations. Therefore, for instance, for vibroisolation of equipment mounted on a symmetric elastic frame, it is expedient to position it in the frame's center of symmetry;
– independence of the various kinds of movements, namely, the lack of any connection between the translational and the rotational displacements as well as between the rotational displacements around the orthogonal axes;

These properties of symmetric systems are well known and can be easily expressed in analytical form.

As has already been pointed out, the group representation theory is a universal method for studying systems with geometric symmetry.

However, its use for solving problems in mechanics has certain specific features (Sect. 4.3) because the mechanical systems have a number of pecularities. They include, in particular, the emergence of a slight asymmetry resulting from the technological disorder of parameters, and the presence in the system of a great number of subsystems possessing different types of symmetry.

L.Y. Banakh, M.L. Kempner, *Vibrations of Mechanical Systems with Regular Structure*, Foundations of Engineering Mechanics, DOI 10.1007/978-3-642-03126-7_4,

This requires certain generalizations regarding the possibility for a successful application of the given theory for mechanical systems. So, for example, we use generalized projective operators for symmetry groups. As a result, it is possible to perform a decomposition of the system in independent subspaces and then study the vibrations as unconnected entities in these subspaces. Such decomposition is performed by means of relatively simple matrix transformations. From a mathematical point of view, this means splitting the initial matrix equations into independent blocks.

Such a method of decomposition reduces the dimensionality of the problem and the parametric spaces, which is particularly important in performing optimization. This approach is particularly convenient because it makes it possible to easily obtain both numerical and analytical results using simple matrix operations. In addition, it permits one to treat complex symmetry types as a series of simpler types with the help of matrix multiplications. This creates the possibility for considering the hierarchy of the symmetry.

4.2 Basic Information about Theory of Groups Representation

One cannot overemphasize the advantages of theory of discrete group representation (or discrete group theory). This approach makes it possible to exhaustively find the main characteristics of the behavior of systems, including dynamic ones, and without solving the motion equation but using symmetry properties alone. In this process, it is easy to identify a number of important properties, such as:

- The number and dimension of the independent unconnected subsystems (invariant subspaces) into which the system is decomposed at free or forced vibrations,
- The multiplicity of the natural frequencies,
- Definition of the structural elements that are not sensitive to the action of these or other loads.

4.2.1 Basic Concepts and Definitions

- If this section appears difficult at first reading, it can be skipped by moving on to the next section, which contains explanations based on sufficiently simple examples.

Let us now introduce some of the basic definitions and concepts of the theory of discrete group representation that will be used later on.

Elements (operations) of symmetry groups and their notations:

1. Rotation at an angle of $2\pi / n$ corresponds to operation C_n and at an angle of $2\pi k / n$ – to operation C_n^k.
2. Reflection in a plane containing a symmetry axis is designated by δ_v and in the plane that is perpendicular to this axis by δ_h.

3. Rotation at an angle of $2\pi / n$ in a plane that is perpendicular to the axis of rotation is designated by S_n.
4. Reflection, i.e., sign change operation, is designated by I.
5. An identical transformation is designated, as usual, by E.

These operations are elements of the given symmetry group G, $g_i \in G$. Every one of these symmetry operations can be represented by a coordinate transformation matrix. Each symmetry operation has its own symmetry operator.

For example, the rotation under an angle of φ around the z axis is described in a matrix form as follows:

$$\boldsymbol{\theta} = \begin{bmatrix} \cos\varphi & -\sin\varphi & \\ \sin\varphi & \cos\varphi & \\ & & 1 \end{bmatrix}.$$

The matrix representation Γ is called a square matrix, which corresponds to the elements of the symmetry group. The trace of the matrix transformation (i.e., the sum of the elements along its major diagonal) is define the character of the group. It is invariant at any change of basic vectors. A representation that has a block-diagonal form is called irreducible. Tables exist showing the characters of different types of symmetry [71].

Let us explain the introduced concepts with the help of sufficiently simple examples. The system in Fig. 4.1a is symmetric with respect to point O.

Masses 1 and 2 can be transformed into one another in the process of the following symmetry operations: E is an identity operation and δ_υ the reflection in the plane containing the symmetry axis. These operations determine symmetry group C_2. The table of characters has the following form:

	E	δ_υ
A	1	1
B	1	–1

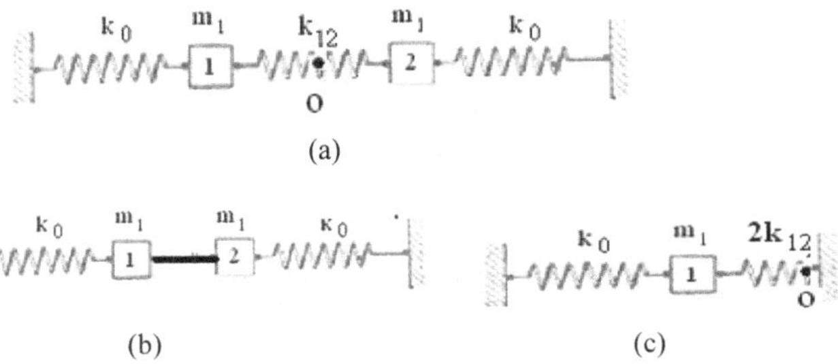

Fig. 4.1a-c A two-mass symmetric system (**a**) and its subsystems, corresponding to symmetrical (**b**) and skew symmetric vibrations modes (**c**)

All symmetry operations (elements) corresponding to the given symmetry type are shown in tables of characters. The coefficient before the symmetry element determines the number of such elements in the group.

Knowing the table of characters it is possible to determine the projective operators for the given symmetry type

$$P_k = \frac{f_k}{n}\sum_i \chi_k(g_i^*)g_i, \tag{4.1}$$

where g_i *is* the elements of group G, χ its character, n is the order of the group, f_k the dimensionality of the kth irreducible representation; $\chi(g^*) = \chi(g^{-1})$ for real operators.

It is known that characters are orthogonal to each other

$$\sum \chi_i^{\mu}\chi_i^{\nu} = \delta_{\mu\nu} = \begin{cases} 0, & \mu \neq \nu \\ 1, & \mu = \nu. \end{cases}$$

It means that any two columns of character table are orthogonal to each other. Therefore $\mathbf{P}^T = \mathbf{P}^{-1}$.

The transformation of coordinates:

$$\mathbf{v} = \mathbf{P}\,\mathbf{x} \tag{4.2}$$

determines the basic (fundamental) vectors. Here **x** is the initial coordinates and **v** the basic vectors in the symmetric coordinates. The initial matrix **D** in the new coordinates **v** can be written with the help of the matrix transformation

$$\mathbf{D}^* = \mathbf{P}^T\mathbf{D}\mathbf{P}.$$

Such matrix transformation follows from the rule of quadratic forms transformation.

With the help of this transformation, we perform a splitting of the initial matrix into a number of independent ones, i.e., we present matrix **K-ΛM**, in block-diagonal form:

$$\mathbf{D}^* = \mathbf{P}^T\mathbf{D}\mathbf{P} = \begin{bmatrix} \mathbf{D}_1 & & & \\ & \mathbf{D}_2 & & \\ & & \ddots & \\ & & & \mathbf{D}_n \end{bmatrix}. \tag{4.3}$$

Each block corresponds to its irreducible representation. The order of the block coincides with the dimensionality of the irreducible representation and, therefore, with the multiplicity of the natural value λ. In this way, the projective operators decompose the initial n-dimensional space into a direct sum of independent invariant subspaces.

Insofar as the basic vectors convert the initial matrix into one of a block-diagonal type, they can be considered as a certain generalization of the vibration modes. In the sequel, we will call them vibration types.

When studying forced vibrations, the vector of the external forces **F** must also be decomposed, in accordance with the symmetry operator, into the form

$$\mathbf{F} = \mathbf{P}\mathbf{F}^*.$$

In this way, if the structure has symmetric properties, its stiffness and inertia matrices can be decomposed with the help of the symmetry operators into submatrices that are independent of each other. This permits one to search for a solution in these independent subspaces. Consequently, one can find results in analytical form and also reduce considerably the dimensionality of the problem since now problems with a considerably lower dimensionality compared with the n-dimensional problem have to be solved.

The application of theory of group representation consists in the following:

1. Determining the symmetry group for the system
2. Determining the character of the irreducible representations for this group with the help of the corresponding tables.
3. Determining the projective operators **V** and the basic vectors for the invariant subspaces.
   More detailed information about theory of discrete group representation can be found, for example, in [38, 61, 71].

4.2.2 Examples of Applying Groups Representation Theory

We will now demonstrate the application of group-theoretic approaches with two simple examples. Let us first consider a two-mass system (Fig. 4.1) with equal masses m_1 and elastic elements k_0, k_{12}. It is obvious that this system has only symmetric and skew-symmetric modes. The symmetric modes $\overline{X}$ correspond to displacements of masses 1 and 2 in phase, the skew-symmetric ones $\overline{\overline{X}}$ – in antiphase; there are no other modes in this system.

An obvious known transformation of coordinates links the initial coordinates X_1, X_2 and $\overline{X}, \overline{\overline{X}}$

$$\overline{X} = \frac{X_1 + X_2}{2}, \ \overline{\overline{X}} = \frac{X_1 - X_2}{2},$$

or, in matrix view,

$$2\begin{bmatrix} \overline{X} \\ \overline{\overline{X}} \end{bmatrix} = \begin{bmatrix} 1 & 1 \\ 1 & -1 \end{bmatrix} \begin{bmatrix} X_1 \\ X_2 \end{bmatrix}. \tag{4.4}$$

The transformation operator is

$$\mathbf{P} = \frac{1}{2}\begin{bmatrix} 1 & 1 \\ 1 & -1 \end{bmatrix}.$$

This operator can be considered as the most simple projective operator of symmetry. It can be obtained also by means of Eq. (4.1).

In the case of symmetric vibration modes, when masses 1 and 2 are moving in phase, the stiffness k_{12} obviously does not work. A system with a mass of $2m_1$ and elasticity of $2k_0$ corresponds to this mode. Such system on Fig. 4.1b is shown. In the case of skew-symmetric modes, the node of vibrations is always in the center of the system o, which is why this point can be considered a fixing and here, therefore, the stiffness of $2k_{12}$ works. Two equal systems with a mass of m_1 and elasticity k_0+2k_{12} correspond to this mode. Such system on Fig. 4.1c is shown. The computational model for these vibration modes can be presented by means of these two systems Fig. 4.1b, c.

Let us now prove these physical considerations in a strictly analytical manner. The equations of the free vibrations for the system in Fig. 4.1a have the form

$$m_1\ddot{X}_1 + (k_0 + k_{12})X_1 - k_{12}X_2 = 0$$
$$m_1\ddot{X}_2 + (k_0 + k_{12})X_2 - k_{12}X_1 = 0,$$

or, in matrix view,

$$\begin{bmatrix} -m_1\lambda^2 + k_0 + k_{12} & -k_{12} \\ -k_{12} & -m_1\lambda^2 + k_0 + k_{12} \end{bmatrix}\begin{bmatrix} X_1 \\ X_2 \end{bmatrix} = \mathbf{DX} = 0.$$

In the new variables (4.4) and in accordance with the rules for the transformation of quadratic forms, this matrix has the view

$$\mathbf{D}^* = \mathbf{P}^{-1}\begin{bmatrix} -m_1\lambda^2 + k_0 + k_{12} & -k_{12} \\ -k_{12} & -m_1\lambda^2 + k_0 + k_{12} \end{bmatrix}\mathbf{P} =$$
$$= 2\begin{bmatrix} -m_1\lambda^2 + k_0 & \\ & -m_1\lambda^2 + k_0 + 2k_{12} \end{bmatrix}.$$

Therefore the initial matrix has been partitioned into two independent sub matrices: The upper sub matrix describes symmetric modes and the lower one – skew-symmetric modes. The following equations correspond to these sub matrices:

$$m_1\ddot{\overline{X}} + k_0\overline{X} = 0$$
$$m_1\ddot{\overline{\overline{X}}} + (k_0 + 2k_{12})\overline{\overline{X}} = 0,$$

which actually corresponds to the two subsystems on Fig. 4.1b and c described above.

In analyzing forced vibrations, all external forces must also be decomposed into symmetric and skew-symmetric ones in accordance with (4.4), which is always

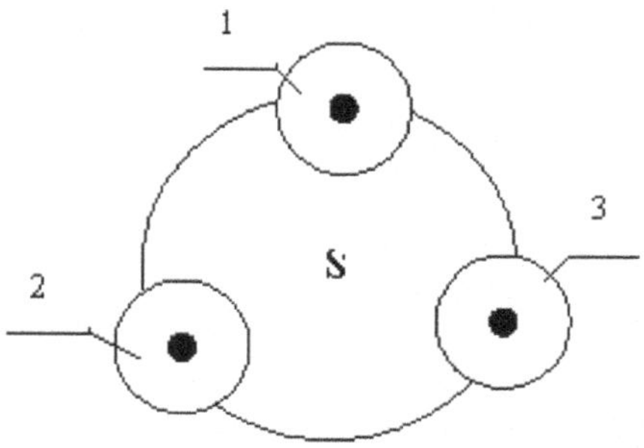

Fig. 4.2 System with cyclic symmetry of type C_{3v}

possible. Then the two subsystems are treated separately in an analogous manner. In this case, it is obvious that if only a symmetrically (or skew-symmetrically) positioned force is acting, then no skew-symmetric (symmetric) vibration modes will be induced. In such a case, it is sufficient to consider only the subsystem in Fig. 4.1b (or Fig. 4.1c).

Let 'we now consider a more complex system presented in Fig. 4.2.

This system has a geometric symmetry of type C_{3v}. Therefore, all its coordinates must change according to this group symmetry. A table for the characters in this group [71]

C_{3v}	E	$2C_3$	3σ
A_1	1	1	1
A_2	1	1	-1
E_1	2	-1	0

The upper row of the table contains the symmetry operators for this group. In our case, these are identical transformation E, two rotations C_3 and $C^{-1}{}_3$ at angles $\pm 2\pi/3$, and three reflections σ_1, σ_2, σ around the axes connecting the corners of the triangle 1, 2, 3 with its center. The characters of the corresponding operators are written in the right-hand part of the table. The left-hand column contains the dimensionality of these characters: A_1, A_2 are one-dimensional subspaces, E_1 is a two-dimensional subspace.

The projective symmetry operators V_i can be determined on the basis of (4.1) using the table of the characters.

$$\begin{aligned} V_1 &= \frac{1}{3}(X_1 + X_2 + X_3) \\ V_2 &\equiv 0 \\ \left.\begin{aligned} V_3 &= \tfrac{1}{3}(2X_1 - X_2 - X_3) \\ V_4 &= \tfrac{1}{3}(-X_1 + 2X_2 - X_3) \end{aligned}\right\} &. \end{aligned} \tag{4.5}$$

In matrix form, this transformation is as follows:

$$\mathbf{V} = \mathbf{P}\,\mathbf{X}, \quad \mathbf{P} = \frac{1}{3}\begin{bmatrix} 1 & 1 & 1 \\ 2 & -1 & -1 \\ -1 & 2 & -1 \end{bmatrix}, \tag{4.6}$$

Where $\mathbf{P}$ is the projective operator in matrix form.

Operators V_i can be treated as vibration modes. The physical meaning of these operators is clear:

- V_1 means that all nodes 1, 2, 3 are in phase;
- V_2, V_3 mean the vibrations corresponding to deformations of the triangle: nodes 1 and 3 are in phase, node 2 is in antiphase.

Operator V_1 is in subspace A_1, $V_2 \equiv 0$ is in subspace A_2, operators V_3, V_4 are in two-dimensional subspace E_1, and, therefore, these operators correspond to multiple roots, and any linear combination of them will also represent an vibration mode.

Note. Determining projective operators with help of Eq. (4.1) is possible for any symmetry type. But in mechanic system there is an often case of cyclic symmetry. The general view of projective operators in this case is given in Sect. 4.4.2.

4.3 Applying Theory of Group Representation to Mechanical Systems: Generalized Projective Operators of Symmetry

4.3.1 Features of Mechanical Systems with Symmetric Structure

As has already been pointed out, the application of the theory of group representation for solving problems in mechanics has some specific features that make them different from the problems in physics and chemistry.

First, in real-life mechanical structures there is almost always a small asymmetry due to technological inaccuracies and assembly errors, so that the system becomes quasi-symmetric. In this case, partition into the independent subspaces or vibration types specified above is incomplete. A weak interaction emerges between them; in addition, some beat also take place due to the mismatch in the multiple frequencies.

Second, symmetry is related to nodes, and each of them in mechanical systems, generally speaking, may be a solid or a whole subsystem with many degrees of freedom.

Third, a mechanical system, as a rule, consists of a number of subsystems, each of which can have its own type of symmetry, so that it is necessary to consider the compatibility of different types symmetries.

In the sequel, mechanical systems usually contain long solids with six degrees of freedom. The question arises as to what symmetry group to allocate the solid to so that its movement could correspond to the symmetry group of the assembled system.

Therefore, a generalization of the available analytical approaches and formation of matrix symmetry operators is necessary.

The calculation methods for symmetric systems are of particular importance in connection with the wide use of the finite-element method (FEM). This also contributes to its features and represents a prospective application of group representation theory in combination with FEM.

In this respect, it is necessary to have available methods for the analysis of both symmetric and quasi-symmetric mechanical systems consisting of various subsystems with many degrees of freedom. These features of mechanical systems required the development of special approaches in group representation theory.

4.3.2 Generalized Projective Operators and Generalized Modes

In order to take into account the above-described characteristics of mechanical systems, generalized projective operators and generalized vibration modes (vibration types) have been introduced.

If each node has only one coordinate, then the operators V_i (4.5) describe the vibration modes. However, elements or subsystems with big dimensionality can be used as nodes in mechanical systems. For example, let us assume that each node in our model in Fig. 4.2 has three degrees of freedom; then the dimensionality of each operator will be equal to 3: it becomes a vector. Therefore, the generalized vibration modes of a symmetric system will be determined by vectors (basic vectors). Henceforth, for simplicity's sake, we will call them *generalized vibration modes* or *vibration types* [2].

The projective operators corresponding to them are *generalized projective operators.*

They can be presented in the form of a block matrix whose elements, related to a node with n degrees of freedom, are matrices of the order (n x n). The generalized operators are determined on the basis of an ordinary symmetry operator by multiplying its elements by the identity matrix **E** of the corresponding order. From a physical point of view, this means that all coordinates in a given node will be vibrates in accordance with the same symmetry operator.

Let us now find the generalized operator **P** for the subsystem in Fig. 4.2. Let each node have 3 degrees of freedom; then the whole system will have 9 degrees of freedom. Taking into account (4.6), the generalized operator has the following form:

$$\mathbf{V} = \mathbf{P}\,\mathbf{X}, \quad \mathbf{P} = \frac{1}{3}\begin{bmatrix} \mathbf{E} & \mathbf{E} & \mathbf{E} \\ 2\mathbf{E} & -\mathbf{E} & -\mathbf{E} \\ -\mathbf{E} & 2\mathbf{E} & -\mathbf{E} \end{bmatrix}. \tag{4.7}$$

Therefore, analogously to (4.5), the projective operator will now determine not scalars but vectors $\mathbf{V}_1, \mathbf{V}_2, \mathbf{V}_3, \mathbf{V}_4$. Their dimension is equal to dimension of corresponding node: in our case it is 3. They lay now in subspaces U_1, U_2, U_3 with the same dimensionality. Analogously to above results vector $\mathbf{V}_1$ lays in subspace U_1 which corresponds to not multiple roots, $\mathbf{V}_2 \equiv 0$ and $\mathbf{V}_3, \mathbf{V}_4$ are in double-dimension subspace U_3 corresponds to multiple roots.

The view of the projective operator (4.7) permits one to draw important conclusions about the frequencies and modes (types) of the system under consideration:

1. There are three nonzero generalized basic vectors (one in the subspace U_1 and two in the double-dimensional subspace(U_3)). The natural frequencies corresponding to the double-dimensional subspace are multiple ones. Therefore, the frequencies of our system (9×9) have three non-multiple and three multiple frequencies.
2. The vibration types (generalized modes) determine the rows of operator (4.7). It can be seen from here that only the following vibration types could exist:

- In-phase vibrations, where all nodes vibrate with the same amplitude [top row in (4.7)]. To these correspond three non-multiple frequencies.
- Antiphase vibrations. These are described by the two last rows in matrix **P**. To these correspond three multiple frequencies. Antiphase vibrations are those in which the neighboring nodes vibrate in antiphase.

We must note that the information about the dynamic properties of the system was obtained without solving the motion equations but only on the basis of analysis of the symmetry groups and the type of projective operators.

Note. Operator (4.7) can be easily converted into a form where all vectors are orthonormalized. In this way it has been taken into account that any linear combination of the natural modes that corresponds to multiple roots is also an natural mode:

$$\mathbf{P} = \begin{bmatrix} \frac{1}{\sqrt{3}} & \frac{1}{\sqrt{3}} & \frac{1}{\sqrt{3}} \\ \frac{2}{\sqrt{6}} & \frac{-1}{\sqrt{6}} & \frac{-1}{\sqrt{6}} \\ 0 & \frac{1}{\sqrt{2}} & \frac{-1}{\sqrt{2}} \end{bmatrix}. \tag{4.7a}$$

Let us now write the stiffness matrix of the system in the new coordinates **V** (4.7)

$$\mathbf{K}^* = \mathbf{P}^T \mathbf{K} \mathbf{P}.$$

It can be easily seen that due to the orthogonality of operator **P**

$$\det \mathbf{K}^* = \det \mathbf{K},$$

and, therefore, their natural frequencies coincide.

The receiving projective operators with help of Eq. (4.1) are suitable for systems with any type of symmetry. However usually, mechanical systems use relatively simple types of symmetry. The general view of projective operators for an n-angular plane frame with cyclic symmetry is given below (Sect. 4.4.2.). Such systems have a *point symmetry* for which at all operators of symmetry one point (the symmetry centre) does not vary.

4.4 Vibrations of Frames with Cyclic Symmetry

4.4.1 Stiffness and Inertia Matrices

Let us now consider the vibrations of an elastic system with cyclical symmetry C_n. Systems of this type are very widely used in engineering. They include symmetric frames, disk turbines with blades, planetary reduction gears, etc.

Let us now consider the plane vibrations of a symmetric elastic n-angular frame with dampers.

A very promising area in the study of mechanical systems is the use of group representation theory in combination with the FEM. Let's create the finite-element model. In this case, however, it is convenient to select a special direction of the coordinate axes in each node, as shown in Fig. 4.3 [12].

Actually, in the use of FEM, the directions of the coordinate axes are traditionally selected to be the same for each node (Fig. 4.3a). Then the stiffness and inertia matrices of the elements included in the given node are written in these axes by rotation of the local coordinate axes of the finite elements at a corresponding angle ϑ(Appendix A). This means multiplication by the matrix of the rotation. However, for convenience of use of the theoretic- group approaches, it is considerably more convenient to select a local coordinate system that is symmetric with respect to each node (Fig. 4.3b).

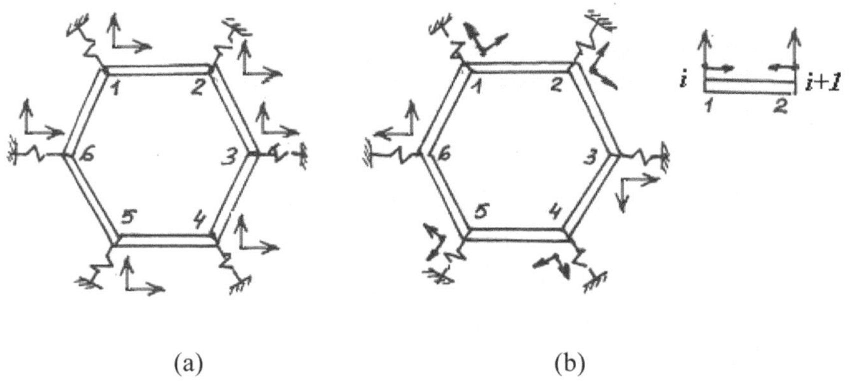

(a) (b)

Fig. 4.3 A symmetric hexagonal frame

Let us assume that the frame is formed by beam elements. For simplicity's sake, let us assume in the beginning that each span of the frame 1-2,..., 6–1 can be represented as an one finite element (although the obtained results will be valid also in the general case as we shall show further).

In these axes – as follows from the algorithm (Appendix C) – the matrix that describes the vibrations takes a particularly simple form, which does not depend on the numbering of the nodes. Thus, for an n-angular frame with a C_{nv}-type symmetry,

$$\begin{array}{l} \text{nodes} \quad 1 \quad 2 \quad 3 \ \ldots \quad n{-}1 \quad n \\ \mathbf{D}_n = \begin{bmatrix} \mathbf{a}_{11} & \mathbf{a}_{12} & 0 & \cdots & 0 & \mathbf{a}_{21} \\ \mathbf{a}_{21} & \mathbf{a}_{11} & 0 & \cdots & 0 & 0 \\ \cdots & \cdots & \cdots & \cdots & \cdots & \cdots \\ \mathbf{a}_{11} & 0 & 0 & \cdots & \mathbf{a}_{21} & \mathbf{a}_{11} \end{bmatrix} - \lambda \mathbf{M} = \begin{bmatrix} \mathbf{d}_{11} & \mathbf{d}_{12} & 0 & \cdots & 0 & \mathbf{d}_{21} \\ \mathbf{d}_{21} & \mathbf{d}_{11} & 0 & \cdots & 0 & 0 \\ \cdots & \cdots & \cdots & \cdots & \cdots & \cdots \\ \mathbf{d}_{11} & 0 & 0 & \cdots & \mathbf{d}_{21} & \mathbf{d}_{11} \end{bmatrix} \end{array} \tag{4.8}$$

where

$$\mathbf{a}_{11} = \boldsymbol{\theta}_{\varphi}^{tr} \mathbf{K}_{11} \boldsymbol{\theta}_{\varphi} + \boldsymbol{\theta}_{-\varphi}^{tr} \mathbf{K}_{22} \boldsymbol{\theta}_{-\varphi} + k$$

$$\mathbf{a_{12}} = \mathbf{a}_{21}^{tr} = \boldsymbol{\theta}_{\varphi}^{tr} \mathbf{K}_{12} \boldsymbol{\theta}_{-\varphi}\,, \quad \mathbf{d}_{ij} = \mathbf{K}_{ij} - \Lambda \mathbf{M}_{ij},$$

$\mathbf{K_n} = \begin{bmatrix} \mathbf{K}_{11} & \mathbf{K}_{12} \\ \mathbf{K}_{21} & \mathbf{K}_{22} \end{bmatrix}$ is the stiffness matrix of the finite element (Chap. 2), $\boldsymbol{\theta}_{\varphi}$ is the matrix of the rotation at an angle $\varphi = 2\pi /n$, and k is the stiffness of the damper.

The matrix structure (4.8) is quite obvious from point of view of symmetry because each node is completely identical in the selected coordinate system.

The stiffness matrix with intermediate nodes positioned on beams 1–2,2–3,..., will be of an analogous type – only its order will be considerably higher.

If we represent the elastic elements as a system with distributed parameters, in this case blocks $\mathbf{D}_{11}, \mathbf{D}_{12} = \mathbf{D}_{21}^{T}$, $\mathbf{D}_{22}$ will be transcendental functions of the frequency; the entire remaining technique, however, remains unchanged.

For simplicity's sake, we will call "symmetric" the coordinates (x, y, z) corresponding to Eq. (4.8) (Fig.4.3b). It can easily be seen that the coordinates (x', y', z') in the traditional coordinate system (Fig. 4.3a) and in the "symmetric" one used by us (Fig. 4.3b) are linked by the relation

$$\mathbf{X}' = \boldsymbol{\theta}\mathbf{X}, \quad 0 = \begin{bmatrix} \mathbf{E} & & & \\ & \boldsymbol{\theta}_{\varphi} & & \\ & & \ddots & \\ & & & \boldsymbol{\theta}_{(n-1)\varphi} \end{bmatrix} \tag{4.9}$$

The decomposition of matrix $\mathbf{D}$, describing the symmetric frame as a result of symmetry takes place irrespective of the system coordinate; but for a matrix like (4.8) this is very easily for analytical studies.

4.4.2 Projective Operators for Frame: Generalized Modes

Initially, we will study the vibrations of an elastic frame with a cyclic symmetry of the hexagonal type. Its stiffness matrix has an analogous form (4.8) at $n = 6$. The given system has a symmetry type C_{6v} and corresponding table of characters [71] is:

C_{6V}	E	C_2	$2C_3$	$2\,C_6$	$3\sigma_v$	$3\,\sigma_d$
A_1	1	1	1	1	1	1
A_2	1	1	1	1	–1	–1
B_1	1	–1	1	–1	1	–1
B_2	1	–1	1	–1	–1	1
E_1	2	–2	–1	1	0	0
E_2	2	2	–1	–1	0	0

Here A_i and B_i are one-dimensional representations: to A_i, correspond in-phase motions, and to B_i antiphase motions; E_1,E_2 are two-dimensional representations with multiple roots.

We will explain the table of characters. As can be seen from this table, the group C_{6v} contains the following elements of symmetry:

- an identity element E,
- 2 C_3 – two rotations at angles $\pm 2\pi / 3$,
- 2 C_6 – two rotations at angles $\pm 2\pi / 6$,
- 3 $\sigma_v, 3\sigma_d$ – three reflections with respect to the axes passing through the corners of the hexagon and the middles of the sides.

The basic (fundamental) vectors $\mathbf{V}_b$, which determine the coordinates transformation in accordance with (4.1), have the view

$$\mathbf{V}_b = \begin{bmatrix} \mathbf{V_1} \\ \mathbf{V_2} \\ \mathbf{V_3} \\ \mathbf{V_4} \\ \mathbf{V_5} \\ \mathbf{V_6} \end{bmatrix} = \frac{1}{6} \begin{bmatrix} 1 & 1 & 1 & 1 & 1 & 1 \\ 2 & 1 & -1 & 2 & -1 & 1 \\ 1 & 2 & 1 & -1 & -2 & -1 \\ 2 & -1 & -1 & 2 & -1 & -1 \\ -1 & 2 & -1 & -1 & 2 & -1 \\ 1 & -1 & 1 & -1 & 1 & -1 \end{bmatrix} \begin{bmatrix} \mathbf{X}'_1 \\ \mathbf{X}'_2 \\ \mathbf{X}'_3 \\ \mathbf{X}'_4 \\ \mathbf{X}'_5 \\ \mathbf{X}'_6 \end{bmatrix} = \mathbf{P}_6\mathbf{X}'. \tag{4.10}$$

Vectors $\mathbf{V}_i$ are located in independent subspaces. Vector $\mathbf{V}_1$ lays in subspace U_1. Vectors $\mathbf{V}_2$ and $\mathbf{V}_3$ are in the double-dimensional subspace U_2 with multiple roots. Vectors $\mathbf{V}_4$ and $\mathbf{V}_5$ are located in the double -dimensional subspace U_3 with multiple roots. Vector $\mathbf{V}_6$ determines the subspace U_4 .

Therefore, if the frame has n natural frequencies, then $n/6$ of them are singlefold – corresponding to the subspace U_1 (vector $\mathbf{V}_1$), $n/3$ are twofold – corresponding to

U_2 (vectors $\mathbf{V}_2$, $\mathbf{V}_3$), $n/3$ are twofold – corresponding to U_3 (vectors $\mathbf{V}_4$, $\mathbf{V}_5$), and $n/6$ are singlefold – corresponding to U_4 (vector $\mathbf{V}_6$).

Vectors located in different subspaces are orthogonal to each other. As regards vectors ($\mathbf{V}_2$, $\mathbf{V}_3$) (and ($\mathbf{V}_4$, $\mathbf{V}_5$)), which are in one subspace, their choice is not uniquely defined because any linear combination of them also belongs to the same subspace. In particular, the cyclic rearrangement of the coordinates makes it possible to obtain new basic vectors $\mathbf{V}'_2, \mathbf{v}'_3$ in the given subspace.

For node with single degree of freedom vectors $\mathbf{V}_i$ became as a scalar V_i and it determines the vibration modes. In the case of multidimensional nodes, $\mathbf{V}_i$ (and therefore matrix $\mathbf{P}_6$) determines the vibration types. In the latter case, matrix $\mathbf{P}_6$ must be presented as a block matrix as $\mathbf{P}_6$ $\mathbf{E}$, where $\mathbf{E}$ is the identity matrix of the corresponding order.

The configuration of the oscillation modes (types) is shown in Fig. 4.4.

The in-phase vibrations with equal displacements of nodes 1–6 correspond to vector $\mathbf{V}_1$. Vectors $\mathbf{V}_2, \mathbf{V}_3$ describe the vibrations of nodes 1-2-3 in antiphase with respect to nodes 4-5-6; this mode (or type) can be considered closed wave sinusoids. The second mode corresponding to the same frequency can be obtained from the first one, for example, by rotating it at an angle of $2\pi/6$. Vectors $\mathbf{V}_4, \mathbf{V}_5$ describe the vibrations of two neighboring nodes in antiphase with the two preceding ones.

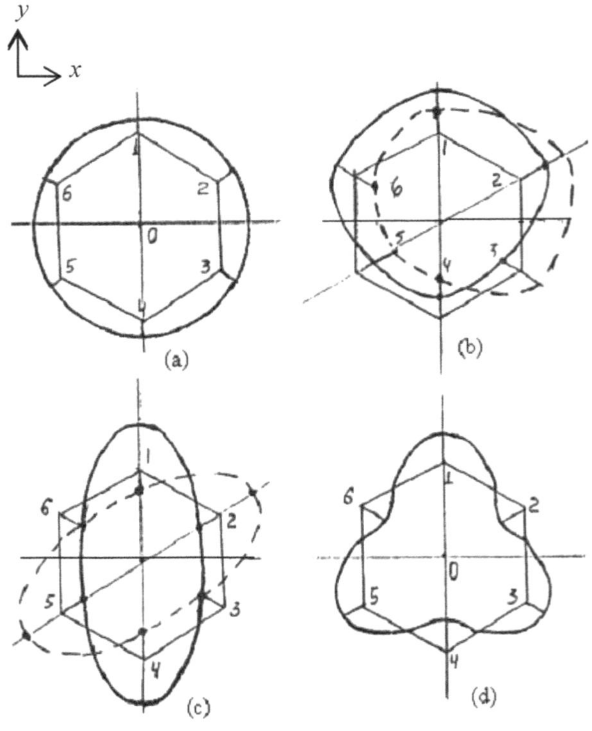

Fig. 4.4 Subspace U_1 (**a**), subspace U_2 (**b**), subspace U_3 (**c**), subspace U_4 (**d**)

Vector $\mathbf{V}_6$ corresponds to the vibrations of all neighboring nodes in antiphase with identical amplitudes (by absolute value). As can be seen from Fig. 4.4, the frame vibrations represent waves with different numbers of nodes.

In this way, if a frame has n natural frequencies, then $n/6$ of them are non-multiple, corresponding to subspace U_1 (vector $\mathbf{V}_1$), $n/3$ are multiple roots, corresponding to subspace U_2 (vectors $\mathbf{V}_2, \mathbf{V}_3$), $n/3$ are multiple roots, corresponding to subspace U_3 (vectors $\mathbf{V}_4, \mathbf{V}_5$), and $n/6$ are non-multiple, corresponding to subspace U_4 (vector $\mathbf{V}_6$).

It can easily be seen that the vectors positioned in different subspaces are orthogonal to each other. As regards the pairs of vectors $(\mathbf{V}_2, \mathbf{V}_3)$ (and $(\mathbf{V}_4, \mathbf{V}_5)$), which are in the same subspace, their choice, as usual, is not uniquely defined.

We again become convinced that the qualitative nature of the symmetric systems vibrations can be obtained without the need to solve the motion equations but by using only the symmetry properties.

In the general case, for cyclical systems with type C_{nv} symmetry, for example in n-angular frame, the operators $\mathbf{P}_n$ (4.1.) obtain the following form:

$$\mathbf{P}_n = \begin{bmatrix} 1 & 1 & \cdots & 1 & 1 \\ \cdots & \cdots & \cdots & \cdots & \cdots \\ \sin q\beta & \cdots & \cdots & \cdots & \sin nq\beta \\ \cos q\beta & \cdots & \cdots & \cdots & \cos nq\beta \\ \cdots & \cdots & \cdots & \cdots & \cdots \\ 1 & -1 & \cdots & 1 & -1 \end{bmatrix} E, \tag{4.11}$$

where $\beta = 2\pi\ /n$, $q = (n - \mathrm{I})\ /2$ for odd n, $q = (n-2)\ /2$ for even n, and $\mathbf{E}$ is the identity matrix whose order is equal to the number of independent coordinates in the node. If the displacement of the node is described by only single coordinate, then $\mathbf{P}_n$ represents the modes of the n-angular frame ,which appear the following:

- The displacements of the nodes have a wavelike nature with wavelengths proportional to 1:2:3. . .n;
- All natural frequencies for plane vibrations (except for the lowest one, and for even n also for the highest one) have a multiplicity equal to two (in the case of spatial vibrations, the multiplicity of the frequencies can be higher);
- The lowest form corresponds to the displacement of all nodes in phase (subspace U_1), and the highest one in antiphase.

In the case of multidimensional nodes, operator $\mathbf{P_n}$ – as has already been pointed out – should be treated as a block one; in this case, it determines the generalized vibration types. The use of vibration types is convenient in the analysis of multidimensional systems where it is impractical to follow up every coordinate and mode individually and desirable to have a more general analog.

However, in order to obtain the further results, it is necessary to use the stiffness and inertia matrices.

Applying the group transformation (4.11) to the matrix (4.8) and setting $n=6$, we obtain block-diagonal matrix $\mathbf{D}^*{}_\mathbf{6}$

$$\mathbf{D}_6^* = \begin{bmatrix} \mathbf{A}_{11}^* & & & & & \\ & \mathbf{A}_{22}^* & \mathbf{A}_{23}^* & & & \\ & \mathbf{A}_{32}^* & \mathbf{A}_{33}^* & & & \\ & & & \mathbf{A}_{44}^* & \mathbf{A}_{45}^* & \\ & & & \mathbf{A}_{54}^* & \mathbf{A}_{55}^* & \\ & & & & & \mathbf{A}_{66}^* \end{bmatrix} - \Lambda \mathbf{M},$$

$$\mathbf{A}_{11}^* = \mathbf{a}_{11} + \mathbf{a}_{12} + \mathbf{a}_{21}, \mathbf{A}_{44}^* = 6\mathbf{a}_{11} - \mathbf{a}_{12} - \mathbf{a}_{21},$$

$$\begin{bmatrix} \mathbf{A}_{22}^* & \mathbf{A}_{23}^* \\ \mathbf{A}_{32}^* & \mathbf{A}_{33}^* \end{bmatrix} = \begin{bmatrix} 2\mathbf{a}_{11} + \mathbf{a}_{12} + \mathbf{a}_{21} & \mathbf{a}_{11} + 2\mathbf{a}_{12} - \mathbf{a}_{21} \\ \mathbf{a}_{11} - \mathbf{a}_{12} + 2\mathbf{a}_{21} & 2\mathbf{a}_{11} + \mathbf{a}_{12} + \mathbf{a}_{21} \end{bmatrix},$$

$$\begin{bmatrix} \mathbf{A}_{33}^* & \mathbf{A}_{34}^* \\ \mathbf{A}_{43}^* & \mathbf{A}_{44}^* \end{bmatrix} = \begin{bmatrix} 2\mathbf{a}_{11} - \mathbf{a}_{12} - \mathbf{a}_{21} & -\mathbf{a}_{11} + 2\mathbf{a}_{12} - \mathbf{a}_{21} \\ -\mathbf{a}_{11} - \mathbf{a}_{12} + 2\mathbf{a}_{21} & 2\mathbf{a}_{11} - \mathbf{a}_{12} - \mathbf{a}_{21} \end{bmatrix}. \tag{4.12}$$

In this way, the initial matrix became decomposed into non-connected blocks corresponding to the invariant subspaces. In order to obtain numerical results, it is sufficient to perform independently calculations for those subsystems whose order does not exceed $n/3$.

For an n-angular frame, after these matrix transformations have been performed, we obtain a block-diagonal matrix in which $(n-1)/2$ (for odd n) and $(n-2)/2$ (for even n) paired diagonal blocks are formed. The decomposition of the matrix takes place independently of the selected coordinate system in the nodes, both in the traditional $\mathbf{X}'$ and in the "symmetric $\mathbf{X}$"; however, the view of matrix (4.8) is more convenient for analytical solutions. The paired blocks correspond to multiple roots are connected between themselves; this fact indicates an interaction between these vibration types. This interaction arises from the following reason: In each node of spatial frame, there are, generally speaking, six independent coordinates – three translational and three rotational ones, each having the same symmetry group C_{nv} and the same operator $\mathbf{P}_n$;. However, due to the phase shifts, their oscillation modes are a linear combination of the basic vector, and in the general case they are not orthogonal to each other. This explains the appearance of the blocks $\mathbf{D_{ij}} \neq 0$. The rows of operator $\mathbf{P_n}$ determine, as usual, the generalized vibration modes.

Note. Frequency expression (2.5) for a regular structure coincides with relationship (4.11) for a system with rotational symmetry C_n. This is easy to understand if one takes into consideration the fact that a system with rotational symmetry is obtained by closing the ends of the system, i.e., under the boundary conditions $\theta(0,t) = \theta(n,t)$. Therefore, the projective operators of symmetry group C_{nv} and groups with translational symmetry must be analogous. The projective operator for a group with translational symmetry for an n-mass system is [25]:

$$
\mathbf{P_{Tr}} = \begin{array}{cccccc}
 & T_0 & T_1 & & \ldots & T_n \\
\Gamma^{(0)} & 1 & 1 & & \ldots & 1 \\
\Gamma^{(1)} & & \exp\left(\frac{2\pi i}{n}\right) & & \ldots & \exp\left(2\pi i\frac{n-1}{n}\right) \\
\vdots & \vdots & \vdots & & \ldots & \vdots \\
\Gamma^{(n-1)} & 1 & \exp\left(2\pi i\frac{n-1}{n}\right) & & \ldots & \exp\left(2\pi i\frac{(n-1)^2}{n}\right)
\end{array}
$$

4.4.3 Analysis of Forced Vibrations

Above described approach is particularly convenient in the analysis of forced vibrations because external forces are usually distributed according to some vibration type. So, the external forces $F(t)$ are transformed according to (4.10):

$$\mathbf{F}^* = \mathbf{P}^T\mathbf{F}.$$

So, for example, if forces with equal size and direction act on each node, then they are distributed according to the first vibration type U_1. In this case, only the first block will remain in the equation describing the forced vibrations:

$$\mathbf{D}_{11}\mathbf{X}_1 = \mathbf{F}^*.$$

4.5 Effect of FE-Mesh on Matrix Structure

A very promising area is the use of group representation theory in combination with the FEM. When FEM is used, the structure is decomposed into equal elements. In this case, it is necessary – just as before – to take into consideration the fact that the transformations of the symmetry groups will be applied to nodes whose dimensionality is greater than 1. Therefore, the transformations must be formed as blocks of matrices of the corresponding order.

Let us assume that each side of the frame is broken down into three finite elements 6–15 (Fig. 4.5).

Additional nodes (6, 7, 8,. . .15) are now introduced on each side 1–2, 2–3. . . of the frame. If the coordinate system is selected as shown in Fig 4.5b, the structure of matrix $\mathbf{K}^{*\mathbf{n}}$ will be analogous to the one provided above (4.8), only its blocks will now have the order (18 × 18) (or, generally speaking, $6q$, where q is the number of finite elements). In particular, block $\mathbf{a}_{11}$ will now be

$$\mathbf{a}_{11} = \begin{bmatrix} d_{11} & d_{12} & \\ d_{21} & b_{11} & b_{12} \\ & b_{21} & b_{22} \end{bmatrix}.$$

Here blocks b_{11} and b_{12} describe the intermediate sections 6–7 and 7–2 and the connections between them; block d_{11} belongs to section 1–6. In this case, operator

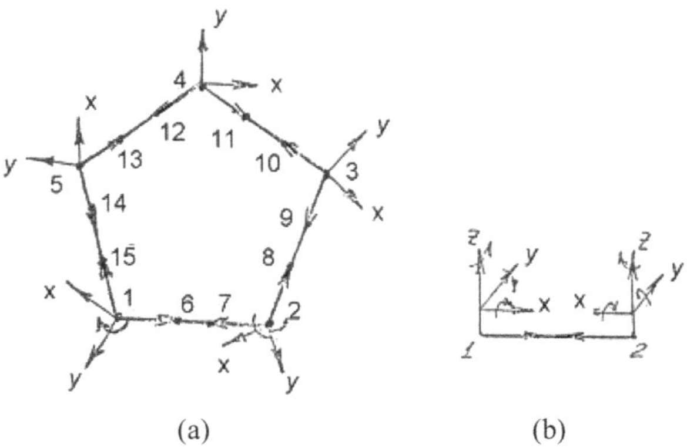

Fig. 4.5a,b Frame decomposed into finite elements. (**a**) sides of frame with finite elements 6–15, (**b**) coordinates system for each finite element

$\mathbf{P}_n$ will have the same structure (4.11) as above, but the dimensionality of the blocks will now be greater. And indeed, the associations of the intermediate nodes 6- 8- 10-12-14 and 7- 9- 11- 13- 15 have a symmetry of the same type as the corners 1- 2- 3-4- 5, and, therefore, they will have a transformation by means of the same operator $\mathbf{P}_n$. Every one of these associations has the same vibration modes (types) as corners 1- 2- 3- 4- 5, but with a certain phase shift. For Fig. 4.5 this shift for the associations will obviously be $2\pi/3$ and $2\pi/6$. And this phase shift causes the appearance of the off-diagonal blocks $d_{12,}\, b_{12}$ that describe the interaction between the vibration modes of these associations.

4.5.1 The Square Frame: Generalized Modes

Let us consider as an example the generalized vibration modes (vibration types) for a square frame. It has symmetry C_{2v} and a projective operator

$$\mathbf{P}_4 = \begin{bmatrix} \mathbf{E} & \mathbf{E} & \mathbf{E} & \mathbf{E} \\ \mathbf{E} & -\mathbf{E} & \mathbf{E} & -\mathbf{E} \\ \mathbf{E} & \mathbf{E} & -\mathbf{E} & -\mathbf{E} \\ \mathbf{E} & -\mathbf{E} & -\mathbf{E} & \mathbf{E} \end{bmatrix},$$

where the first row corresponds to invariant subspace U_1, the second one to subspace U_2, and the third and fourth rows to the double-dimensional subspaces U_3.

It can be seen from this that the following generalized modes (vibration types) exist. In subspace U_1, all equally located nodes are in phase (Fig. 4.6a). In subspace U_2, all equally located neighboring nodes are in antiphase (Fig. 4.6b). Multiple roots correspond to the double-dimensional subspace U_3. Their vibration types are as follows: two neighboring and equally located nodes in phase and two others in antiphase with respect to them (Fig. 4.6c, d).

It is necessary to point out a very important property: The vibration types obtained on the basis of the generalized operator $\mathbf{P_n}$ *are invariants* for the given symmetry type. They do not depend on the number of finite elements; moreover, they are preserved if we consider a system with distributed parameters. All vibration modes of the system are in these four vibration types. To each type corresponds its own series of natural frequencies.

These vibration types do not change when any changes take place in the parameters and even in the structure of the system if the specified symmetry type does not change. They are *system invariants*. The presence of such invariants determines the good predictability of the work, the convenience of the calculations, and the optimization of the structure.

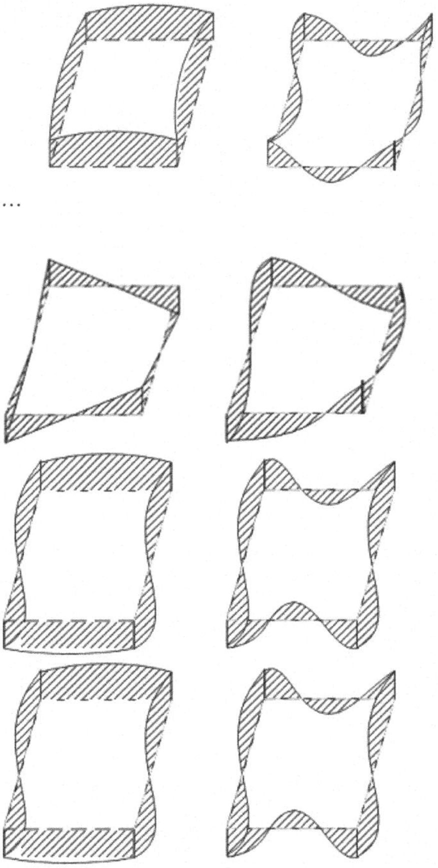

(**a**)First vibration type (all identical nodes are in phase) (Subspace U_1)

(**b**) Second vibration type (all identical nodes are in antiphase)(Subspace U_2)

(**c,d**) Second and third vibration types. Two neighboring identical nodes are in phase, the two others in antiphase with respect to them (Double-dimensional subspace U_3)

Fig. 4.6a-d Generalized vibration modes (vibration types) for a square frame

4.6 Vibroisolation of Body on Symmetrical Frame: Vibrations Interaction

The symmetry operators are particularly convenient for analyzing the vibrations interactions in a system. We will demonstrate this using as an example the vibroisolation of a solid body mounted on an elastic frame. For simplicity's sake, let us consider a plane system of vibroisolation [in the (x, y) plane] of the body m_7 mounted in the center of symmetry of an elastic hexagonal frame (Fig. 4.7). The body has three degrees of freedom (x, y, φ).

Initially, let us find the independent vibration types that occur in a system by its symmetry and then we establish the interaction between the vibrations of the frame and the body m_7. Let us construct the stiffness matrix of this system. The stiffness matrix $\mathbf{K}_7$ of this system has the form

node1 2 ... 6 7

$$\mathbf{K}_7 = \begin{bmatrix} \mathbf{a}_{11} & \mathbf{a}_{12} & \cdots & \mathbf{a}_{21} & -\mathbf{K}_{am}\boldsymbol{\theta}_{\varphi 1} \\ & \mathbf{a}_{11} & \cdots & \cdots & -\mathbf{K}_{am}\boldsymbol{\theta}_{\varphi 1} \\ & & \ddots & & \vdots \\ \text{symm.} & & & \mathbf{a}_{11} & -\mathbf{K}_{am}\boldsymbol{\theta}_{\varphi 16} \\ & & & & \sum_i \boldsymbol{\theta}_{\varphi_i}^T \mathbf{K}_{am}\boldsymbol{\theta}_{\varphi_i} \end{bmatrix},$$

$$\mathbf{K}_{am} = \begin{bmatrix} K_x & & \\ & K_y & \\ & & K_\theta \end{bmatrix}$$

is the sum of the stiffness of the inside and outside dampers.

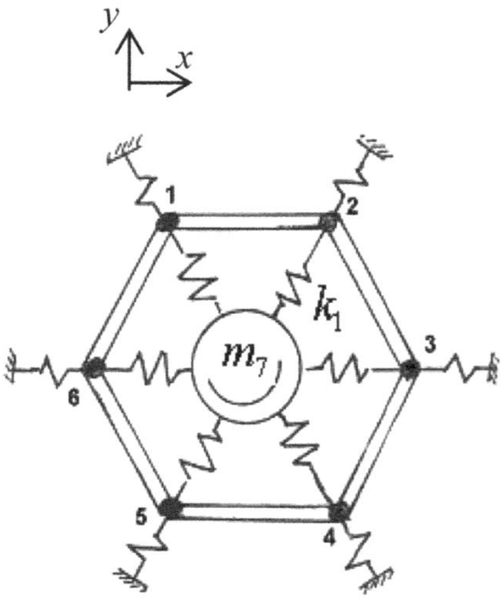

Fig. 4.7. Vibroisolation of a body m_7 mounted in the center of symmetry of an elastic frame

Let us now form the symmetry operator

$$\mathbf{P}_7 = \begin{bmatrix} \mathbf{P}_6 & \\ & \mathbf{E} \end{bmatrix}.$$

Here $\mathbf{P}_6$ was determined earlier in (4.10), the identity block $\mathbf{E}$ corresponds to the body m_7 and indicates that no symmetry transformation is applied to it. By performing the matrix multiplication, we establish that matrix $\mathbf{K}_7^*$ has the following structure:

$$\mathbf{K}^*{}_7 = \mathbf{P}_7{}^T\mathbf{K}_7\mathbf{P}_7 =$$
$$\begin{bmatrix} \mathbf{A}_{11}^* + \mathbf{K}^*{}_{am} & & & & & & \mathbf{A}_{17}^* \\ & \mathbf{A}_{22}^* & \mathbf{A}_{23}^* & & & & \mathbf{A}_{27}^* \\ & \mathbf{A}_{32}^* & \mathbf{A}_{33}^* & & & & \mathbf{A}_{37}^* \\ & & & \mathbf{A}_{44}^* & \mathbf{A}_{45}^* & & \\ & & & \mathbf{A}_{54}^* & \mathbf{A}_{55}^* & & \\ & \text{symm.} & & & & \mathbf{A}_{66}^* & \\ & & & & & & \mathbf{A}_{77}^* \end{bmatrix}.$$

Here blocks $\mathbf{A}_{ij}^*$ $(i,j = 1...6)$ were determined above in (4.12). Blocks $\mathbf{A}_{i7}^*$ $(i = 1, 2, 3)$ determine the interaction between the frame and the body.

In order to analyze the dynamic interaction in the system, it is sufficient to analyze only the structure of matrix $\mathbf{K}^*{}_7$. The block in the bottom right corresponds to the mass m_7. It follows from this that the vibrations of the mass m_7 are related only with subspaces U_1 and U_2, i.e., to the displacements and rotations of the frame. At the same time, the vibration types in U_3 reflect the deformation of the frame itself and are not related with the body vibrations.

For a more detailed analysis, we will now find the actual expressions for the blocks of the obtained matrix. They have the form

$$\mathbf{A}_{17}^* = \begin{bmatrix} 0 & 0 & 0 \\ 0 & 0 & 0 \\ 0 & 0 & K_\varphi \end{bmatrix}, \quad \begin{bmatrix} \mathbf{A}_{27}^* \\ \\ \mathbf{A}_{37}^* \end{bmatrix} = \begin{bmatrix} -K_x\frac{\sqrt{3}}{2} & K_x/2 & 0 \\ -K_x/2 & -K_x\frac{\sqrt{3}}{2} & 0 \\ 0 & 0 & 0 \\ -K_y\frac{\sqrt{3}}{2} & -K_y/2 & 0 \\ K_y/2 & -K_y\frac{\sqrt{3}}{2} & 0 \\ 0 & 0 & 0 \end{bmatrix}$$

From this we see that the angular and transversal body vibrations are not related to each other. The angular vibrations are related to subspace U_1 and the translational ones to U_2.

Taking into account the actual expressions for $\mathbf{A}_{11}^*$, another independent mode vibration of frame can be determined. This is the frame's rotation as a rigid body around an axis passing through the center of symmetry perpendicularly to the plane of the figure. And indeed, $\mathbf{A}_{11}^*$ have the form

$$\mathbf{a}_{11} + \mathbf{a}_{12} + \mathbf{a}_{21} = \begin{bmatrix} \frac{48EJ}{\ell^3}\sin^2\varphi + K_x & 0 & 24\frac{EJ}{\ell^2}\sin\varphi \\ & 4\frac{EF}{\ell}\sin^2\varphi + Ky & 0 \\ 24\frac{EJ}{\ell^2}\sin\varphi & 0 & \frac{12EJ}{\ell} \end{bmatrix}.$$

We can see that the second diagonal term is not related to other elements of the system. This term corresponds to vibrations over the y axis. It is included in U_1, and because of that it describes vibrations of the frame that have identical shifts along the y axis. In the accepted coordinate system (Fig. 4.3), this means rotation of the entire frame around the symmetry axis. Let us recall that the initial and the transformed ("symmetric") coordinates are connected by relations (4.9) and (4.10).

In this way, as can be seen from the provided examples, the analysis by means of the structure of the transformed stiffness matrix $\mathbf{K}_7^*$ makes it possible to obtain sufficiently complete qualitative information about the dynamic interactions of a system.

4.7 Quasi-symmetrical Systems

As it has already been pointed out, the real-life mechanical structures do not contain strictly symmetric systems and almost always there is slight asymmetry in them. We will call quasi-symmetric this type of systems. It is natural to raise the point about the vibrations in quasy-symmetric systems having in mind that the strictly symmetric system has multiple frequencies.

4.7.1 Vibrations Interaction at Slight Asymmetry

As an example, let us consider the vibrations of an elastic square frame on dampers (Fig. 4.8). Frames of this kind are often used as platforms under power generation equipment (turbine rotors, etc.) and also as load-bearing structures. The vibrations of the frame have a considerable effect on the dynamics of an assembled system. As a rule, it is practically impossible to make the frame strictly symmetrical: there is always discrepancy of the parameters, which leads to interactions between the vibrations and to the emergence of beats caused by mismatches in the multiple frequencies.

Let us now consider the vibration interactions in the square frame as a result of the existence of a slight symmetry in the stiffness of dampers 1,2,3,4. Let us assume that the discrepancy of the stiffness parameters of the dampers is small and is equal to Δ_i, i.e., the stiffness of the ith damper: $\mathbf{K}_d + \Delta_i$.

The matrix describing the vibrations in this case is

$$\mathbf{D}_4 = \begin{bmatrix} \mathbf{a}_{11} + \Delta_1 & \mathbf{a}_{12} & & \mathbf{a}_{21} \\ & \mathbf{a}_{11} + \Delta_2 & \mathbf{a}_{12} & \\ & & \mathbf{a}_{11} + \Delta_3 & \mathbf{a}_{12} \\ \text{symm.} & & & \mathbf{a}_{11} + \Delta_4 \end{bmatrix} - \Lambda\mathbf{M} = 0. \qquad (4.13)$$

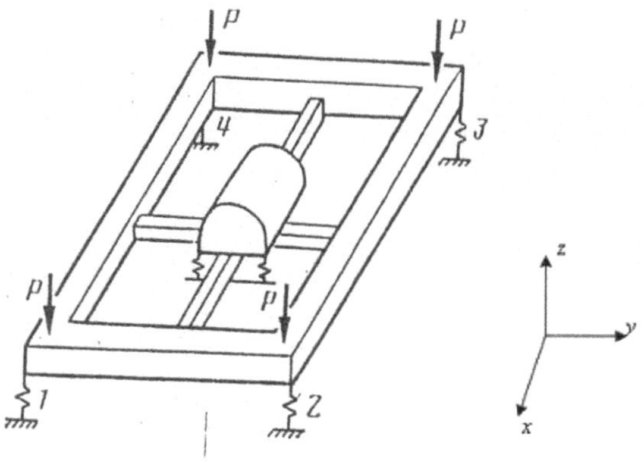

Fig. 4.8 A square frame on dampers with a slight discrepancy of the stiffness parameters

The basic vectors of the symmetry operator for a square frame have been determined above:

$$\begin{aligned}
\mathbf{V}_1 &= 1/4(\mathbf{X}_1 + \mathbf{X}_2 + \mathbf{X}_3 + \mathbf{X}_4) - \text{subspace } U_1 \\
\mathbf{V}_2 &= 1/4(\mathbf{X}_1 - \mathbf{X}_2 + \mathbf{X}_3 - \mathbf{X}_4) - \text{subspace } U_2 \\
\mathbf{V}_3 &= 1/4(\mathbf{X}_1 + \mathbf{X}_2 - \mathbf{X}_3 - \mathbf{X}_4) - \text{subspace } U_3 \\
\mathbf{V}_4 &= 1/4(\mathbf{X}_1 - \mathbf{X}_2 - \mathbf{X}_3 + \mathbf{X}_4) \quad \text{subspace } U'_3,
\end{aligned}$$

where $\mathbf{X}_i$ is the vector of displacement for the ith node. Subspaces (U_3 U'_3) generate a two-dimensional subspace, so that multiple frequencies correspond to the basic vectors.

The symmetry operator is

$$\mathbf{P}_4 = \begin{bmatrix} \mathbf{E} & \mathbf{E} & \mathbf{E} & \mathbf{E} \\ \mathbf{E} & -\mathbf{E} & \mathbf{E} & -\mathbf{E} \\ \mathbf{E} & \mathbf{E} & -\mathbf{E} & -\mathbf{E} \\ \mathbf{E} & -\mathbf{E} & -\mathbf{E} & \mathbf{E} \end{bmatrix},$$

Applying symmetry transformation $\mathbf{P}_4$ to matrix $\mathbf{D}_4$ (4.13), we obtain

$$\mathbf{D}_4^* = \begin{bmatrix} \mathbf{A}_{11}^* & \mathbf{\Delta}_{12} & \mathbf{\Delta}_{13} & \mathbf{\Delta}_{14} \\ & \mathbf{A}_{22}^* & \mathbf{\Delta}_{21} & \mathbf{\Delta}_{23} \\ & & \mathbf{A}_{33}^* & \mathbf{\Delta}_{34} \\ \text{symm.} & & & \mathbf{A}_{44}^* \end{bmatrix} - \Lambda \mathbf{M}^* = 0, \tag{4.14}$$

$$\mathbf{A}^*_{ii} = \mathbf{A}_{ii} + \frac{1}{4}\sum \mathbf{\Delta}_i,$$

$$\mathbf{\Delta}_{12} = \frac{1}{4}(\mathbf{\Delta}_1 - \mathbf{\Delta}_2 + \mathbf{\Delta}_3 - \mathbf{\Delta}_2), \quad \mathbf{\Delta}_{14} = \mathbf{\Delta}_{34} = \frac{1}{4}(\mathbf{\Delta}_1 - \mathbf{\Delta}_2 - \mathbf{\Delta}_3 + \mathbf{\Delta}_2)$$

$$\mathbf{\Delta}_{13} = \mathbf{\Delta}_{21} = \mathbf{\Delta}_{23} = \frac{1}{4}(\mathbf{\Delta}_1 + \mathbf{\Delta}_2 - \mathbf{\Delta}_3 - \mathbf{\Delta}_2).$$

As can be seen from Eqs. (4.14) and (4.14a), in the case of distribution of the asymmetry over U_2, i.e., if $\Delta_1 = -\Delta_2 = \Delta_3 = -\Delta_4$, links arise between blocks $\mathbf{A}^*_{11}$ and $\mathbf{A}^*_{22}$, i.e., between subspaces U_1 and U_2. Links between blocks $\mathbf{A}^*_{33}$ and $\mathbf{A}^*_{44}$, i.e., inside the subspaces U_3 and U'_3 also arise. From here it is easy to draw the following conclusions:

- If the external force is distributed over U_1, i.e., $F_1 = F_2 = F_3 = F_4$, then the emergence of resonance interactions at the natural frequencies belonging to subspaces U_1 and U_2 is possible. In this case, no inclination of the frame around the x and y axes take place.
- If the external force is distributed over U_3, i.e., $F_1 = F_2 = -F_3 = -F_4$, then no vibrations will take place in subspaces U_1 and U_2. In this case, there will be no translational displacements: the center of the frame will remain fixed.

Analogous conclusions can also be obtained for other kinds of distribution of the asymmetry and the external forces.

Therefore, asymmetry has a significant effect on the dynamics of a system. The translational vibration types along the x and y axes and rotation about these axis in this case are not independent; and beats caused by mismatches of the natural frequencies emerge. This means that if we place the rotor in the geometrical center of the frame, good vibroisolation may not be achieved. Nonetheless, if we know the discrepancy of the dampers parameters and also the distribution of the external load, we can position the dampers in such a way that their asymmetry does not cause interactions between the undesired vibration types, thus ensuring a good vibroisolation of the object.

4.7.2 Quasi-symmetrical Systems: Free Vibrations

Let us now study some patterns of the free and forced vibrations of quasi-symmetric systems. Outside of the main diagonal in matrix $\mathbf{D}_4$ (4.13) are small terms that are proportional to the discrepancy level of parameters Δ_i. Therefore, the matrix describing the vibrations in quasi-symmetric systems can be written in a form containing a small parameter $\varepsilon \ll 1$

$$\mathbf{D}^* = \\ = \begin{bmatrix} \mathbf{D}_{11} + \varepsilon(\mathbf{B}_{11} - \Lambda\mathbf{Z}_{11}) & \mathbf{D}_{12} + \varepsilon(\mathbf{B}_{12} - \Lambda\mathbf{Z}_{12}) & \cdots & \mathbf{D}_{1n} + \varepsilon(\mathbf{B}_{1n} - \Lambda\mathbf{Z}_{1n}) \\ & & \cdots & \mathbf{D}_{2n} + \varepsilon(\mathbf{B}_{2n} - \Lambda\mathbf{Z}_{2n}) \\ \text{symm.} & \cdots & \cdots & \cdots \\ & & & \mathbf{D}_{nn} + \varepsilon(\mathbf{B}_{nn} - \Lambda\mathbf{Z}_{nn}) \end{bmatrix}, \tag{4.15}$$

or

$$\mathbf{D}^* = \mathbf{D}_0 + \boldsymbol{\varepsilon}(\mathbf{B} - \boldsymbol{\lambda}\mathbf{Z}),$$

where $\mathbf{D}_{ii} = \mathbf{K}_{ii} - \Lambda\mathbf{M}_{ii}$, $\mathbf{D}_{ij} = \mathbf{K}_{ij} - \Lambda\mathbf{M}_{ij}$ describes the generating strictly symmetrical system. Matrices $\mathbf{B}_{ij}$ describes the asymmetry of the elastic elements and $\mathbf{Z}_{ij}$ – of the inertia elements.

The structure of this matrix coincides with the matrices of the weakly-interacted systems described in Chap. 2 (Sect. 2.8). For $\varepsilon = 0$ we obtain a generating system corresponding to a system with ideal symmetry. To evaluate the interaction level, it is necessary to use the dimensionless coefficients of dynamic interconnection

$$\alpha_{ij}^{pr} = \frac{(\mathbf{h}_i^p)^T \mathbf{K}_{ij} \mathbf{h}_j^r}{\left(\omega_i^{(p)} \omega_j^{(r)}\right)}, \quad p = 1...n_i, \; r = 1...n_j. \tag{2.31}$$

This coefficient reflects the ratio between the work performed by the connections between subsystems (or coordinates) i and j and the energy of these partial systems.

If in matrix (4.13) of square frame there is the single coordinate in node only then its terms became as scalar. And in this case coefficients (2.31) became

$$\alpha_{ij} = \frac{\Delta_{ij}}{\sqrt{\alpha_{ii}\alpha_{jj}}}$$

The methods of the perturbation theory are used for systems described by Eq. (4.15). It was shown in Sect. 2.9 that the solution for weakly-interconnected systems depends essentially on the presence of multiple roots in the generating system. Obviously, the multiple roots occurs in cases where among the symmetry operators there are two-dimensional (or, generally speaking, r-dimensional) representations. For a square frame, these are subspaces U_3 and U'_3 with matrices $\mathbf{A}_{33}^*$ and $\mathbf{A}_{44}^*$ corresponding to them.

If the matrix of the perturbing terms $\varepsilon(\mathbf{B} - \Lambda\mathbf{Z})$ contains only elements that connects the different subspaces without multiple roots, then the corrections both for the frequencies and the normal vectors will be of the order of ε:

$$\begin{aligned} \mathbf{\Lambda} &= \mathbf{\Lambda}_0 + \varepsilon^2 \mathbf{\Lambda}^{(1)} + \ldots, \\ \mathbf{H} &= \mathbf{H}_0 + \varepsilon \mathbf{H_0 S} + \ldots \end{aligned} ,$$

where $\mathbf{\Lambda}_0, \mathbf{H}_0$ is the matrix of the eigenvalues and the eigenvectors of the generating symmetric system at $\varepsilon = 0$. Corrections to them $\Delta_1\lambda_1$ and S_{ij} are determined in Sect. 2.8

But if the perturbation matrix contains elements $\varepsilon\,(\mathbf{B}_{i,\,i+1} - \lambda\,\mathbf{L}_{i,\,i+1})$ connecting the double-dimensional subspaces U_r and U'_r, then we have a case of multiple roots. In this case, the normal vectors of the perturbed matrix is a linear combination of the corresponding normal vectors of the generating one:

$$\mathbf{h}_i = \mathbf{h}_i \pm \mathbf{h}_j + \varepsilon s_{ij} \mathbf{h}_i + o(\varepsilon). \tag{2.34}$$

In this way, at multiple frequencies, the strong interactions between the vibrations corresponding to these frequencies occurs; in addition, a beat mode caused by the slight mismatch of the natural frequencies ω_r and ω_{r+1} occurs.

4.7.3 Quasi-symmetrical Systems: Forced Vibrations

The equations for the forced vibrations of a quasi-symmetric system written in coordinates $\mathbf{V}_i$ are as follows:

$$\mathbf{D}_\varepsilon \mathbf{V} = (\mathbf{D}_0 + \varepsilon \mathbf{D}')\mathbf{V} + \mathbf{F}(t),$$

or

$$\begin{bmatrix} \mathbf{D}_{11}+\varepsilon(\mathbf{B}_{11} - \Lambda\mathbf{Z}_{11}) & \mathbf{D}_{12}+\varepsilon(\mathbf{B}_{12}-\Lambda\mathbf{Z}_{12}) & \cdots & \mathbf{D}_{1n}+\varepsilon(\mathbf{B}_{1n} - \Lambda\mathbf{Z}_{1n}) \\ & & \cdots & \mathbf{D}_{2n}+\varepsilon(\mathbf{B}_{2n} - \Lambda\mathbf{Z}_{2n}) \\ \text{symm.} & \cdots & \cdots & \cdots \\ & & & \mathbf{D}_{nn}+\varepsilon(\mathbf{B}_{nn} - \Lambda\mathbf{Z}_{nn}) \end{bmatrix} \begin{bmatrix} \mathbf{V}_1 \\ \mathbf{V}_2 \\ \vdots \\ \mathbf{V}_n \end{bmatrix} =$$

$$= \begin{bmatrix} \mathbf{F}_1^*(t) \\ \mathbf{F}_2^*(t) \\ \vdots \\ \mathbf{F}_n(t) \end{bmatrix}. \tag{4.16}$$

The damping is included in the terms containing ε.

At a strict symmetry system, i.e., at $\mathbf{B}_{ij} = \mathbf{Z}_{ij} = 0$, each of the forces F^*_i causes a displacements only in its own subspace U_r, and the distribution of the amplitudes has the same symmetry group.

However, in the presence of asymmetry, interaction between different vibration types takes place, and there is spread of the displacements and strains in the structure, which often leads to undesired phenomena. Obviously, the relative value of this interaction depends on the form of asymmetry, and the positioning of the external forces $\mathbf{F}^*_i$. Let us find the conditions for the occurrence of a maximum level of interaction.

Following the perturbation theory, solution (4.16) becomes

$$\mathbf{V} = \mathbf{V}_0 + \varepsilon \mathbf{V} + \cdots,$$

where $\mathbf{V_0}$ corresponds to the vibrations of a system with ideal symmetry. By substituting this solution into (4.16) and equating the terms at identical powers of ε, we obtain

$$\mathbf{V} = \mathbf{V}_0 + \mathbf{D}_0^{-1}\mathbf{D}_1\mathbf{V}_0 + \cdots \tag{4.17}$$

Obviously, the convergence condition of series (4.17) has the following form:

$$\left\| \mathbf{D}_0^{-1} \right\| \, \|\mathbf{D}_1\| < 1, \tag{4.18}$$

where $\| \mathbf{D} \|$ is norm of matrix $\mathbf{D}$.

Let us assume for definiteness that only one force $F_r{}^*$ distributed according to some subspace U_r is acting on the system. Let the frame asymmetry after its expansion over the symmetry operators connects subspace U_r with subspaces U_{r1},U_{r2}, and let U_{r2} be, in turn, connected to U_j.

Then it is sufficient to consider only a fragment of this matrix which describes the vibrations caused by this force:

$$\begin{bmatrix} \mathbf{D}_j & & & \varepsilon\mathbf{B}_{j\,r2} \\ & \mathbf{D}_r & \varepsilon\mathbf{B}_{r\,r1} & \varepsilon\mathbf{B}_{r\,r2} \\ \text{symm.} & & \mathbf{D}_{r1} & \\ & & & \mathbf{D}_{r2} \end{bmatrix} \begin{bmatrix} \mathbf{V}_j \\ \mathbf{V}_r \\ \mathbf{V}_{r1} \\ \mathbf{V}_{r2} \end{bmatrix} = \begin{bmatrix} 0 \\ \mathbf{F}_r^* \\ 0 \\ 0 \end{bmatrix}. \tag{4.19}$$

Based on the above assumption, the remaining blocks of the matrix are not connected with the given block and no internal forces are acting on them. As analysis (4.19) shows, the forced vibrations occur only in subspaces U_r,U_{r1}, U_{r2}, U_j; there is no contribution from vibrations in other subspaces. We will show that the corrections from connection with U_j are of the order of ε^2. And indeed, we obtain from Eqs. (4.17) and (4.19):

$$\begin{bmatrix} \mathbf{V}_j \\ \mathbf{V}_r \\ \mathbf{V}_{r1} \\ \mathbf{V}_{r2} \end{bmatrix} = \begin{bmatrix} 0 \\ \mathbf{V}_{r0} \\ 0 \\ 0 \end{bmatrix} + \begin{bmatrix} 0 \\ 0 \\ \mathbf{V}_{r0}\mathbf{D}_{r1}^{-1}\mathbf{B}_{r\,r1} \\ \mathbf{V}_{r0}\mathbf{D}_{r2}^{-1}\mathbf{B}_{r\,r2} \end{bmatrix} + o(\varepsilon^2). \tag{4.20}$$

It follows from Eq. (4.20) that if the frequency of the exciting force F_r does not cause resonance conditions in any of the subspaces U_{r1}, U_{r2}, U_j, then the corrections to coordinates $\mathbf{V}_{r1}$ and $\mathbf{V}_{r2}$ are determined by the third summand in (4.20). Therefore, the value of these corrections is of the order of ε^2 and series (4.20) converges.

But in the resonance case, the frequency of the exciting force coincides with the natural frequency in one of the subspaces U_{r1}, U_{r2} where resonance vibrations with great amplitudes arise. Then the third summand in (4.20) is very large, condition (4.18) is not fulfilled, and series (4.20) diverges. A dangerous kind of asymmetry can be determined in each concrete case.

4.8 Hierarchy of Symmetries: Multiplication of Symmetries

It often happens in mechanical systems that the subsystems as such have an own type of symmetry, an "internal" symmetry that, generally, does not coincide with the symmetry type of the assembled system. In this way, a symmetry hierarchy emerges (Figs.4.9 and 4.10). As an example of such systems, we can mention, for example, planetary reduction gear (Sect. 4.10), where each of the satellites has a symmetry

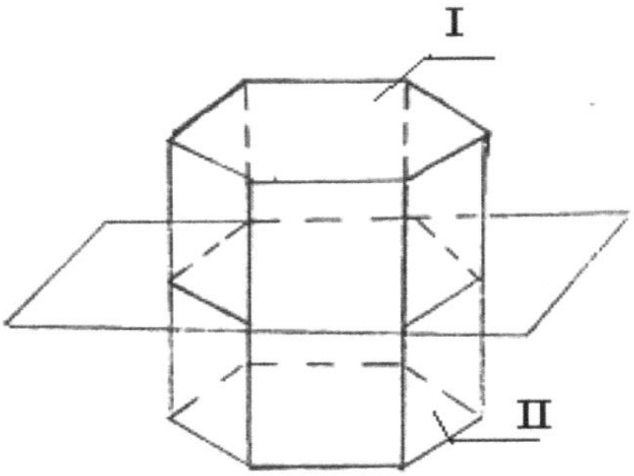

Fig. 4.9 Prism-type symmetry

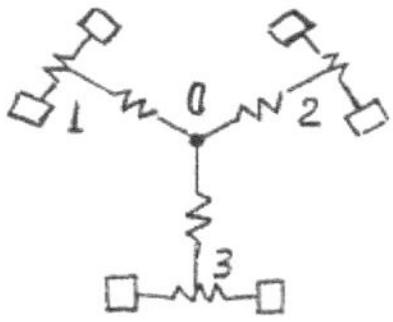

Fig. 4.10 System with a symmetry ierarchy

of the type C_{2v} and n satellites along with transmission gear and an epicycle – symmetry C_{nv}.

Another simplest example is a system with symmetry of the prism type (Fig. 4.9). Planes I and II have a type C_{2v} symmetry, and planes I and II as such have a type C_{nv} symmetry. Let us study the features of such a system and find the corresponding calculation algorithms.

We will note that the subsystems (Fig. 4.9) or subsystems 1,2,3 (Fig. 4.10) have a symmetry of the n-angular type – $C_{nv.}$ Projective operator of symmetry is $\mathbf{P}_n$. Coordinates transformation in planes I and II:

$$\mathbf{V} = \mathbf{P}_n\mathbf{X}. \tag{4.21}$$

However, here there is additional symmetry between planes I and II on Fig. 4.9 or between the identical masses in nodes 1,2,3 (Fig. 4.10). This additional symmetry is of the C_{2v} type and the operator of symmetry $\mathbf{P}_2$ corresponds to it. Then, applying this operator to coordinates $\mathbf{V}$ (4.21), we obtain the resultant symmetry operator as a multiplication of the previous ones:

$$\mathbf{V}' = \mathbf{P}_2\mathbf{V} = \mathbf{P}_n\mathbf{P}_2\mathbf{X}. \tag{4.22}$$

This equation can be considered a multiplication of symmetries. In this way, it is possible to obtain a hierarchy of the dynamic models detailing each time the structure of the subsystems.

For Fig. 4.10, the product of the symmetries $C_{nv} \oplus C_{2v}$ gives the symmetry C_{6v} for the hexagon.

Although these results have been obtained for rather simple systems, they are nonetheless correct for more general cases of vibratory systems as well.

4.9 Periodic Systems Consisting from Symmetrical Elements

Structures consisting of regularly located identical symmetric cells are often used in mechanical engineering; this includes honeycomb structures, catenary's structures, various kinds of roof structures, vaults, etc.

The combination of the wave approach with methods of group theory has proven effective in the study of similar systems. The application of these methods permits one to split the spectrum of a system and to obtain independent branches of the dispersion curve corresponding to independent vibration types of a given symmetry group. This makes it possible to separate the low-frequency and the high-frequency branches. The first branches describe the vibration of a system without deformation of the cells; the following, in contrast, describe their deformation. Then the resulting vibration represent a superposition of waves: a low-frequency wavelength and internal waves of the symmetric cells.

At the numerical level, this means decomposition of the initial problem, which has $N = qnT$ degrees of freedom (where n is the number of cells, q the number of nodes in a cell, T the number of DOF in the node), into q equations on the order of T (and well conditioned at that).

For exampe, for a plane cover system consisting of squares the problem is reduced to solving four third-order equations. Let us write down the view of this solution.

The finite-difference equation describing the rth cell is

$$\mathbf{D}_{rr+1}Z_{r\,r+1} + \mathbf{D}_{r\,r-1}Z_{rr-1} + \mathbf{D}_{r\,r}Z_{rr} + \mathbf{D}_{r+1r}Z_{r+1r} + \mathbf{D}_{r-1\;r}Z_{r-1\,r} = 0, \quad (4.23)$$

where $\mathbf{D}_{r\,r}$ is the matrix describing the cell as a partial subsystem and $\mathbf{D}_{r\,r-1,\dots}$ the connection matrices.

Assuming, as in Chaps. 3,6 $Z_{rj} = x_r y_j$, we obtain a dispersion equation with an order equal to the number of degrees of freedom of a cell.

For simplicity's sake, let us assume that in each cell the coordinates system has been selected in the same manner, as in Sects. 4.4 and 4.7. Then, as a result of the symmetry:

$$\mathbf{D}_{r\,r-1} = \mathbf{D}_{r-1\;r} = \mathbf{D}_{r+1\,r} = \mathbf{D}_{r\,r+1}.$$

And the dispersion equation from (4.23) is

$$\mathbf{D}_{r\,r-1}\,(ch\mu + ch\nu) + \mathbf{D}_{r\,r} = 0.$$

Taking into account the symmetry of the square cell, we can apply operator $\mathbf{P_4}$ to it. We obtain the following equation:

$$\mathbf{P}_4^T[\mathbf{D}_{r\,r-1}\,(ch\mu + ch\nu) + \mathbf{D}_{r\,r}]\mathbf{P}_4 = 0.$$

Then matrix $\mathbf{D}_{rr}$ is obviously decomposed into four blocks corresponding to the independent vibration types as on Fig. 4.6. As a result, we obtain four branches of the dispersion equation. The own series of natural frequencies corresponds to each branch. A number of examples for such mechanical real-life systems with reflection symmetry elements and antisymmetric elements are given in Part II.

These approaches are widely used also in molecular spectroscopy [89], where many branches with a corresponding spectrum are normally used.

4.10 Generalized Modes in Planetary Reduction Gear due to Its Symmetry

It is known that the operation of planetary reduction gear is accompanied by vibrations of its elements such as differential gear, epicycle, and satellites. This worsens considerably the quality of its operation and can lead to its curvature and damaging. There are many studies devoted to the numerical calculation of the dynamic characteristics of the planetary reduction gear [1]. Here, however, we present the *analytical* approaches for the dynamic analysis of this mechanism.

The planetary reduction gear has a high level of geometric symmetry, and this property permits one to apply the theory of groups representation. To this end, it is necessary, above all, to have a dynamic model reflecting the symmetry of the system taking into consideration the connections between its elements.

4.10.1 Dynamic Model of Planetary Reduction Gear

A model of the step of the planetary reduction gear is presented in Fig. 4.11 [4, 13, 43]. The step consists of a differential gear S (its mass, moment of inertia, and radius are equal to m_1, J_1, r_1, respectively). It engages with three satellites St_i (i=1,2,3) (their masses and radii are equal to each other and to m_2, J_2, r_2). For their part, the satellites engage the epicycle Ep (its parameters are m_3, J_3, r_3); they are fixed on a carrier on elastic supports with stiffness h_6. The stiffness of the teeth in the engagement between the differential gear and the satellites is equal to h_1, in the engagement between the epicycle and the satellites are h_3, γ is the engagement angle. A simplified model is considered in which the central wheels and satellites are represented as absolutely rigid bodies; the effect of their weight is not taken into consideration. Such idealization is possible in the study of low-frequency vibrations. We neglect the deformations of the teeth because their effect on the free vibrations in the low-frequency range is limited.

For simplicity's sake, let us consider the plane vibrations of the planetary reduction gear: transversal (x, y) and angular (φ) vibrations (without casing). The stiffness matrix can be presented in block form:

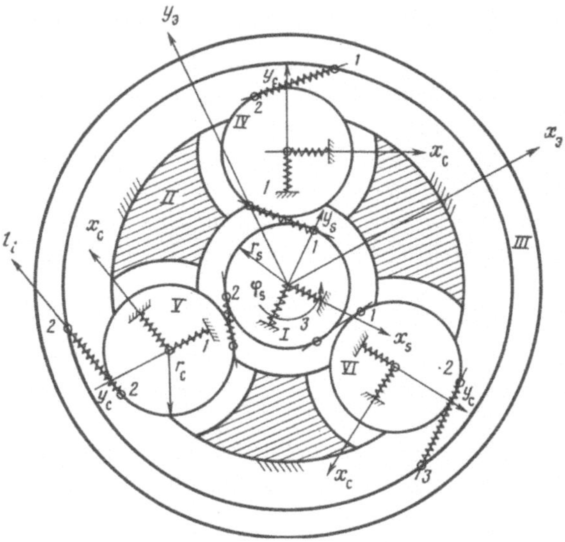

Fig. 4.11 Model of planetary reduction gear. S: differential gear; Ep: epicycle; 1,2,3: satellites (St_i)

$$\mathbf{K} = \begin{bmatrix} \mathbf{K}_S & & \mathbf{K}_{SSt_1} & \mathbf{K}_{SSt_2} & \mathbf{K}_{SSt_3} \\ & \mathbf{K}_{Ep} & \mathbf{K}_{EpSt_1} & \mathbf{K}_{EpSt_2} & \mathbf{K}_{EpSt_3} \\ & & \mathbf{K}_{St_1} & & \\ & symm & & \mathbf{K}_{St_2} & \\ & & & & \mathbf{K}_{St_3} \end{bmatrix}. \quad (4.24)$$

Here, the blocks of the stiffness matrix (3 × 3) for the corresponding subsystems are on the main diagonal, and the blocks of the stiffness matrix describing the connections between these subsystems are outside of the main diagonal. The actual form of these blocks is presented in Appendix D.

In this way, there are 15 generalized coordinates:

$$\mathbf{X} = (\mathbf{X}_S, \mathbf{X}_{\mathrm{Ep}}, \mathbf{X}_{\mathrm{St1}}, \mathbf{X}_{\mathrm{St2}}, \mathbf{X}_{\mathrm{St3}}) = x_{\mathrm{S}}, y_{\mathrm{S}}, \varphi_{\mathrm{S}};\ \ x_{\mathrm{Ep}}, y_{\mathrm{Ep}}, \varphi_{\mathrm{Ep}}, (x_{\mathrm{St}_1}, y_{\mathrm{St}_1}, \varphi_{\mathrm{St}_1} \ldots \varphi_{\mathrm{St}_3}).$$

Therefore, the order of matrix $\mathbf{K}$ is (15 × 15); the inertia matrix $\mathbf{M}$ (15 × 15) is diagonal.

The equations of the free vibrations are

$$\mathbf{M}\ddot{\mathbf{X}} + \mathbf{K}\mathbf{X} = \mathbf{0}.$$

4.10.2 Generalized Normal Modes in Planetary Reduction Gear: Decomposition of Stiffness Matrix

Let us consider the vibrations of the subsystems of a reduction gear [14].

4.10.2.1 Vibration Types in Satellite Subsystems

The positioning of the satellites in a planetary reduction gear has a geometric symmetry of the type C_{3v}. Therefore, all coordinates of the satellites will also change according to this group symmetry. This subsystem has the form shown in Fig. 4.2, and its free vibrations were analyzed above in Sect. 4.2 with the help of the symmetry operator $P_3 = P_{\mathrm{st}}$.

Let us now consider an assembled system of a planetary reduction gear. The projective operator **P** for it can be represented as a block-diagonal matrix

$$\mathbf{P} = \begin{bmatrix} \mathbf{E} & & \\ & \mathbf{E} & \\ & & \mathbf{P}_{\mathbf{St}} \end{bmatrix}, \mathbf{E} = \begin{bmatrix} 1 & & \\ & 1 & \\ & & 1 \end{bmatrix}. \tag{4.25}$$

Each diagonal block in **P** is related, respectively, to the differential gear S and the epicycle Ep and also to the satellites St_i (i =1, 2, 3). The identity matrices **E** relate to the differential gear and the epicycle; it means that no coordinates transformation is applied to them and their coordinates are not transformed. Applying transformation **P** to the stiffness matrix of the reduction gear (4.24), we obtain

$$\mathbf{K}^* = (P)_T(\mathbf{K})(P) =$$

$$= \begin{bmatrix} \mathbf{K}_{\mathrm{S}} & & \mathbf{K}_{\mathrm{SSt}_1} & \mathbf{K}_{\mathrm{SSt}_2} & \mathbf{K}_{\mathrm{SSt}_3} \\ & \mathbf{K}_{\mathrm{Ep}} & \mathbf{K}_{\mathrm{EpSt}_1} & \mathbf{K}_{\mathrm{EpSt}_2} & \mathbf{K}_{\mathrm{EpSt}_3} \\ & & \mathbf{K}_{\mathrm{St}_1} & & \\ & \text{symm.} & & \mathbf{K}_{\mathrm{St}_2} & \\ & & & & \mathbf{K}_{\mathrm{St}3} \end{bmatrix} \tag{4.26}$$

$$\Rightarrow \mathbf{K}^* = \begin{bmatrix} \mathbf{K}_{\mathbf{I}}^* & & 0 \\ & \mathbf{K}_{\mathbf{II}}^{*(1)} & \\ 0 & & \mathbf{K}_{\mathbf{II}}^{*(2)} \end{bmatrix},$$

where

$$\mathbf{K}_{\mathbf{I}}^* = \left[\begin{array}{cc|ccc} a_2 & & \sqrt{3}r_1h_1\cos\gamma & -\sqrt{3}r_1h_1\sin\gamma & \sqrt{3}r_1r_3h_1 \\ & b_2 & \sqrt{3}r_3h_2\cos\gamma & \sqrt{3}r_3h_2\sin\gamma & -\sqrt{3}r_2r_3h_2 \\ - & - & - & - & - \\ & & & & \\ & & & \mathbf{K}_{\mathrm{st}} & \\ & & & & \end{array}\right], \tag{4.27a}$$

This matrix describes the angular coordinates of the differential gear and the epicycle, as well as the coordinates of the satellites, in subspace U_1, i.e., the coordinates $\left(\varphi_{\mathrm{S}}, \varphi_{\mathrm{Ep}}, \mathbf{V}_{1\mathrm{st}}\right)$.

The matrix

$$\mathbf{K}_{\mathbf{II}}^{*(1)} = \mathbf{K}_{\mathbf{II}}^{*(2)} =$$

$$\left[\begin{array}{cc|ccc} a_1 & & -\sqrt{3/2}h_1\cos\gamma & \sqrt{3/2}h_1\sin\gamma & -\sqrt{3/2}r_1h_1 \\ & b_1 & -\sqrt{3/2}h_3\cos\gamma & -\sqrt{3/2}h_3\sin\gamma & \sqrt{3/2}r_3h_3 \\ - & - & - & - & - \\ & & & & \\ & & & \mathbf{K}_{st} & \\ & & & & \end{array}\right] \tag{4.27b}$$

describes the coordinates $(x_S,\ x_{Ep},\ \mathbf{V}_{st3}),\ (y_S,\ y_{Ep},\ \mathbf{V}_{st4})$.

In this way, the initial matrix **K** (15 × 15) is decomposed into three independent blocks (5 × 5) describing independent vibration types in the coordinates $\mathbf{V}_i$. The inertia matrix **M** remains diagonal due to the orthogonal nature of matrix **P**; that is why the independent vibration types is determined only by the stiffness matrix $\mathbf{K}^*$. The generalized coordinates corresponding to (4.26) are as follows:

$$\mathbf{X}^* = (\mathbf{X}_S,\ \mathbf{X}_{Ep},\ \mathbf{V}_{St1},\ \mathbf{V}_{St3},\ \mathbf{V}_{St4}).$$

4.10.3 Free Vibrations

As the analysis of matrix (4.26) shows, the decomposition of the stiffness matrix and, therefore, the decomposition of the vibration types and the spaces of the parameters are determined only by the symmetry properties of the system. Submatrices $\mathbf{K}_{\mathbf{I}}^*$ and $\mathbf{K}_{\mathbf{II}}^{*(1)} = \mathbf{K}_{\mathbf{II}}^{*(2)}$ (4.27a, b) describe the following independent vibration types:

- First vibration type [subspace I – submatrix $\mathbf{K}_{\mathbf{I}}^*$ (4.27a)]: in- phased angular vibrations of the differential gear and the epicycle + in-phase vibrations of the satellites. The dimensionality of this subspace is equal to 5. The main parameters are

$$r_1, r_2, r_3, h_1, h_3, h_6, h_{12}, r_{13}, h_9.$$

- Second type [subspace II – submatrix $\mathbf{K}_{\mathbf{II}}^{*(1)} = \mathbf{K}_{\mathbf{II}}^{*(2)}$ (4.27b)]: transversal vibrations of the differential gear and the epicycle + antiphase vibrations of the satellites. This matrix is decomposed into two equal blocks $\mathbf{K}_{\mathbf{II}}^{*(1)} = \mathbf{K}_{\mathbf{II}}^{*(2)}$ (5 × 5), which indicates the presence of five multiple frequencies in the system.

In this way, by taking into consideration only the symmetry properties, it is possible to perform a sufficiently deep analysis of the dynamic properties of the systems of planetary reduction gear. In addition, the process of optimization of the system can be simplified considerably since the space of the parameters is also divided.

The obtained decomposition of the vibration types has a general character and remains valid for any changes in the system parameters that do not disorder its symmetry type.

The further motions decomposition is possible only by introducing an additional symmetry into the system, which is possible with a special selection of the parameters only.

4.10.4 Forced Vibrations due to Slight Error in Engagement

Let us now consider the vibrations interactions arising at an error in the engagement. Figure 4.12 shows the exciting forces emerging at error Δ_1 in the engagement differential gear satellites; similar forces arise when there is error Δ_2 in the engagement epicycle satellites [43].

Then the vector of the external forces in the reduction gear is

$$\mathbf{F}^T = \begin{bmatrix} 0 \; 0 \; 3h_1 r_1 \Delta_1 \;\; 0 \; 0 \; 3h_3 r_3 \Delta_2 \;\; f_{\mathrm{st}1} \; f_{\mathrm{st}1} \; \mu_{\mathrm{st}1} \cdots \mu_{\mathrm{st}3} \end{bmatrix}$$
$$f_{\mathrm{st}1} = f_{\mathrm{st}2} = f_{\mathrm{st}3} = -h_1\Delta_1 - h_3\Delta_2, \quad \mu_{\mathrm{st}1} = \mu_{\mathrm{st}2} = \mu_{\mathrm{st}3} = h_1 r_2 \Delta_1 - h_3 r_2 \Delta_2.$$

We will show that in this case the vector of the exciting forces has zero elements in subspace II. And indeed, by applying the projective operator (4.25) to vector **F**, we obtain as a result the following equation for the forced vibrations of the reduction gear:

$$\mathbf{K}^* = \begin{bmatrix} \mathbf{K}_{\mathbf{I}}^* & & \\ & \mathbf{K}_{\mathbf{II}}^{*(1)} & \\ & & \mathbf{K}_{\mathbf{II}}^{*(2)} \end{bmatrix} - \Lambda \mathrm{diag}\mathbf{M} = \begin{bmatrix} \mathbf{F}_{\mathbf{I}}^* \\ 0 \\ 0 \end{bmatrix},$$
$$\mathbf{F}_{\mathbf{I}}^{*tr} = \begin{bmatrix} 3h_1 r_1 \Delta_1 \; 3h_2 r_2 \Delta_2 \; -3h_1\Delta_1 - 3h_3\Delta_2 \; -3h_1\Delta_1 - 3h_3\Delta_2 \; 3h_1 r_2 \Delta_1 - 3h_1 r_2 \Delta_2 \end{bmatrix}.$$

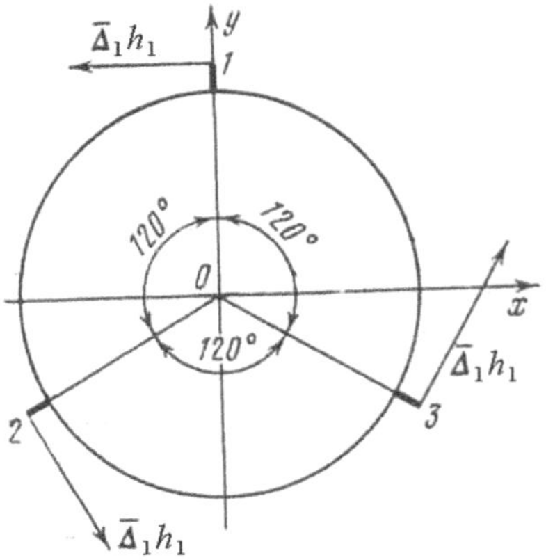

Fig. 4.12 Exciting forces arising at error Δ_1 in the engagement differential satellites

It follows from this that vibrations take place only in subspace I, i.e., there are angular vibrations of the differential gear and the epicycle + vibrations of the satellites in a phase.

In this way, the engagement error does not cause transversal vibrations in the central wheels.

The character of the vibrations can be analyzed analogously when the satellites have the same eccentricity, which induces their vibrations along the x and y axes. By applying symmetry transformation (4.25), it is easy to show that in this case, on the contrary, the external forces act only in subspace II and, therefore, cause transversal vibrations of the differential gear + vibrations of the satellites in antiphase.

4.10.5 Vibrations Interaction at Violation of Symmetry

As a rule, due to errors in the process of manufacturing and assembling real-life mechanical structures, a certain asymmetry occurs in a system, so that the system becomes "quasi-symmetric." Let us assume that one of the satellites (for example, the second one) has a manufacturing error, so that its stiffness differs slightly by a value of δ, i.e., it is necessary to set in stiffness matrix (4.24) $\mathbf{K}_{\mathrm{St}_1} = \mathbf{K}_{\mathrm{St}_3}, \quad \mathbf{K}_{\mathrm{St}_2} + \delta$.

By applying symmetry operator (4.6) to the subsystem of the satellites, we obtain

$$\tilde{\mathbf{K}}_{st}^{*} = \begin{bmatrix} \mathbf{K}_{\mathrm{st}_1} & & \\ & \mathbf{K}_{\mathrm{st}_2} & \\ & & \mathbf{K}_{\mathrm{st}_3} \end{bmatrix} + \Delta\mathbf{K}_{\mathrm{st}}, \Delta\mathbf{K}_{\mathrm{st}} = \delta \begin{bmatrix} 1/3 & 2/\sqrt{18} & -1/\sqrt{6} \\ 2/\sqrt{18} & 2/3 & -2/\sqrt{2} \\ -1/\sqrt{6} & -2/\sqrt{2} & 1/2 \end{bmatrix}. \tag{4.28}$$

It can be seen from this that now the stiffness matrix for the subsystems of the satellites in the $\mathbf{V}_i$ coordinates is no longer diagonal, as was the case with identical satellites. This means that all vibrations of the satellites – both angular and transversal – become interconnected.

Let us now analyze this asymmetry effect on the vibrations of the assembled planetary reduction gear. Here it is necessary to take into account also the occurrence of asymmetry in the engagement satellites – differential gear δ_{SSt_2}, satellites – epicycle δ_{EpSt_2}.

$$\mathbf{K}_{\mathrm{SSt}_2} = \mathbf{K}_{\mathrm{SSt}_1} + \delta_{\mathrm{SSt}_2}, \ \mathbf{K}_{\mathrm{SSt}_3} = \mathbf{K}_{\mathrm{SSt}_1} \qquad \mathbf{K}_{\mathrm{SSt}_2} = \mathbf{K}_{\mathrm{SSt}_1} + \delta_{\mathrm{SSt}_2}, \ \mathbf{K}_{\mathrm{SSt}_3} = \mathbf{K}_{\mathrm{SSt}_1}.$$

After taking into account these errors, stiffness matrix $\tilde{\mathbf{K}}$ takes the following form:

$$\tilde{\mathbf{K}} = \mathbf{K} + \Delta\mathbf{K}, \quad \Delta\mathbf{K} = \left[\begin{array}{cc|c} & & \delta_{\mathrm{SSt}_2} \\ & & \delta_{\mathrm{EpSt}_2} \\ \hline & & \\ \delta_{\mathrm{SSt}_2} & \delta_{\mathrm{EpSt}_2} & \delta \\ & & \end{array}\right] \tag{4.29}$$

Applying symmetry operator (4.25), we obtain

$$\tilde{\mathbf{K}}^{*} = \mathbf{P}^{T}(\mathbf{K} + \Delta\mathbf{K})\mathbf{P} = \begin{bmatrix} \mathbf{K}_{\mathrm{I}}^{*} & & \\ & \mathbf{K}_{\mathrm{II}}^{*(1)} & \\ & & \mathbf{K}_{\mathrm{II}}^{*(2)} \end{bmatrix} + \Delta\mathbf{K}^{*};$$

$$\Delta\,\mathbf{K}^{*} = \left[\begin{array}{c|ccc} & \delta_{\mathrm{SSt}_2} & \delta_{\mathrm{SSt}_2} & \delta_{\mathrm{SSt}_2} \\ & \delta_{\mathrm{EpSt}_2} & \delta_{\mathrm{EpSt}_2} & \delta_{\mathrm{EpSt}_2} \\ \hline & & & \\ \text{symm.} & & \tilde{\mathbf{K}}_{St}^{*} & \\ & & & \end{array}\right]. \tag{4.30}$$

$\tilde{\mathbf{K}}_{\mathrm{St}}^{*}$ is defined in (4.28). As can be seen, all oscillation types in the reduction gear happened to be connected. After taking into account the small value of the error, we can conclude that the vibrations of the subsystems in (4.30) are weakly connected. It is possible to evaluate quantitatively the level of interaction of the vibrations using the concept of dimensionless coefficient of weak energy interactions α_{ij} (2.31).

It is clear from the above discussion that it is possible to compensate the asymmetry in a system with the help of external forces only if these forces are applied to all elements of the system.

Part II
Systems with Distributed Parameters

Foreword to Part II

In the Part II of given book, as against a Part I, as an elementary cell, elements with the distributed parameters with the known analytical solution (beams, plates, shells) are used. For them dynamic compliances (stiffness) are determined. The further analytic and numerical investigations will be carried out with use of fundamental matrixes for regular systems (Chap. 5).

Note. It is necessary to keep in mind, as we use the conceptions of dynamic compliances (stiffness) then as follows from its definition (Sect. 2.3) the all parameters include the current frequency ω. Therefore it should be pointed out that everywhere in Part II we understand by the concepts of *(generalized) forces and coordinates the amplitude of these parameters in the case of harmonic excitation.*

Chapter 5
Basic Equations and Numerical Methods

5.1 Elementary Cells: Connectedness

This chapter contains the basic approaches and equations that are to be used later on for vibration analysis in different kinds of regular structures. These methods are based on the application of the dynamic compliances (stiffness) methods.

Figure 5.1 shows an elementary cell of a mechanical oscillation system. This cell is typical for our later investigations: It has only one connection channel to the neighboring cells (input and output). Therefore, this system is not ramified. However connection channel is, generally speaking, multidimensional.

Here $\mathbf{Q}_{as}, \mathbf{Q}_{bs}$ are the vectors of the internal generalized forces and $\mathbf{q}_{as}, \mathbf{q}_{bs}$ are the vectors of the generalized displacements.

The rule of signs for displacements and forces corresponds to the selected coordinate system (e.g., [31]). The internal forces are positive at the input to the cell if their direction is opposite to the direction of the corresponding positive displacements. The internal forces are positive at the output from the cell if their direction coincides with the direction of the corresponding positive displacements.

We will call the left-hand cross section "input" and we will assign a subscript index "a" to all quantities related to this cross section. And, analogously, we will call the right-hand cross section "output" and we will assign a subscript index "b" to all quantities related to this cross section.

In terms of wave theory, we can call such a cell a generalized one-dimensional geometric grid.

System cells. Connectedness. Let us consider a system of sequentially connected cells. The cells of the system can be two- or three-dimensional, or, generally speaking, l-dimensional. We will assign to each cell a whole number $k(s = 1, 2 \ldots .n)$, n

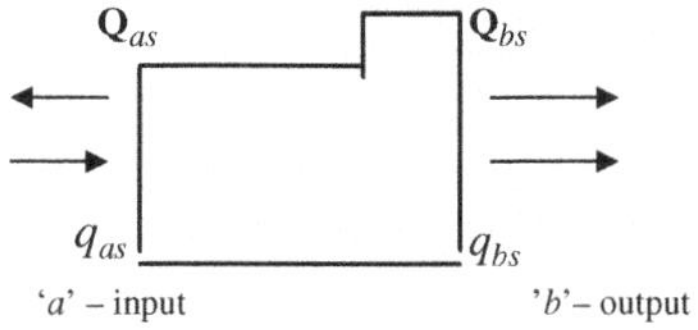

Fig. 5.1 Cell in a mechanical oscillation system

L.Y. Banakh, M.L. Kempner, *Vibrations of Mechanical Systems with Regular Structure*, Foundations of Engineering Mechanics, DOI 10.1007/978-3-642-03126-7_5,

is the number of cells in the system. For example, when studying the simultaneous vibrations of turbine blades connected by a shroud, it is convenient to select as a cell one blade along with a section of the shroud. When studying the shaft vibrations with bladed disks, a disk with a section of the shaft between two disks can be selected. In the case of a rotor with a drum structure, a disk with a section of the conical shell between two disks can be selected as a cell.

Let us now consider two consecutive cells s and $s+1$. Let the connection between these cells be implemented over a cross section S. We will call a connectedness in this cross section the number of independent generalized coordinates that determine the position of this cross section during oscillations. For example, for a spatial rod, the displacement of each cross section is determined by six coordinates. If we have r rods between two cells, then the connectedness of the system in this cross section is equal to $6r$. The connectedness can be different in different cross sections of the system. For a crankshaft on rigid supports, the connectedness in any cross section is equal to 6, however it is equal to 4 in a cross section above the support [52]. In a system of blades with a ring connection, the connectedness in a cross section of the shroud is equal to 6. However, in a cross section where the shroud joins the blade, the connectedness is equal to 5 [54].

When studying the oscillations of a system, it is always recommended to divide the system into cells with a minimal connectedness between them.

5.2 Fundamental Matrices for Systems with Regular Structure

Let us now provide the fundamental matrices we will use constantly in our further investigations. We will find the basic relations between the matrices of dynamic stiffness and compliances and the transition matrix for a system of cells.

5.2.1 Matrices of Dynamic Compliance and Dynamic Stiffness

In this section, we will consider the relationships between the forces and displacement vectors at the input and the output of the cells.

Let the following hold at the input to a cell:

l: a connectedness in the cross section of the cell, $\mathbf{q}_a$: an l-dimensional vector of the generalized coordinates (displacements), $\mathbf{Q}_a$: a vector of the generalized internal forces.

Analogously, at the output from the cell:

t: a connectedness in the cross section of the cell, $\mathbf{q}_b$: a t–dimensional vector of the generalized coordinates (displacements), $\mathbf{Q}_b$: a vector of the generalized internal forces.

Vectors $\mathbf{q}_a$ and $\mathbf{q}_b$ completely determine the position of the input and output cross sections during vibrations.

Now we can write the following matrix equality:

$$\begin{bmatrix} \mathbf{q}_a \\ \mathbf{q}_b \end{bmatrix} = \begin{bmatrix} \mathbf{e}_{aa}\mathbf{e}_{b\alpha} \\ \mathbf{e}_{ba}\mathbf{e}_{bb} \end{bmatrix} \begin{bmatrix} -\mathbf{Q}_a \\ \mathbf{Q}_b \end{bmatrix}, \tag{5.1}$$

where $\mathbf{e}_{aa}$ is the input matrix of dynamic compliance,

$\mathbf{e}_{bb}$ is the output matrix of dynamic compliance,

$\mathbf{e}_{ba}$, $\mathbf{e}_{ab}$ are, respectively, the collateral matrices of dynamic compliance.

Here $\mathbf{e}_{aa}$ and $\mathbf{e}_{bb}$ are square matrices of the orders $(l \times l)$ and $(t \times t)$, respectively, and $\mathbf{e}_{ba}$ and $\mathbf{e}_{ab}$ are rectangular matrices of the orders $(l \times t)$ and $(t \times l)$, respectively.

With a minus sign before $\mathbf{Q}_a$, this matrix becomes symmetric.

We will call the matrix

$$\mathbf{e} = \begin{bmatrix} \mathbf{e}_{aa} \ \mathbf{e}_{ab} \\ \mathbf{e}_{ba} \ \mathbf{e}_{bb} \end{bmatrix} \tag{5.2}$$

a fundamental matrix of dynamic compliance (MDC) of the cell.

Based on the Rayleigh principle, matrices $\mathbf{e}_{aa}$ and $\mathbf{e}_{bb}$ are symmetrical. Matrices $\mathbf{e}_{ba}$ and $\mathbf{e}_{ab}$ are mutually transposed.

$$\mathbf{e}_{ba} = \mathbf{e}_{ab}{}^{T}. \tag{5.3}$$

The matrix of dynamic stiffness (or rigidity) (MDS) is the inverse to the matrix of dynamic compliances (MDC) (for those values of ω at which MDC is not singular)

$$\mathbf{R} = \mathbf{e}^{-1} \tag{5.4}$$

where

$$\mathbf{R} = \begin{bmatrix} \mathbf{R}_{aa}\mathbf{R}_{ab} \\ \mathbf{R}_{ba}\mathbf{R}_{bb} \end{bmatrix}, \tag{5.5}$$

$\mathbf{R}$ is the fundamental matrix of dynamic stiffness of the cell, $\mathbf{R}_{aa}$, $\mathbf{R}_{ab}$, $\mathbf{R}_{ba}$, $\mathbf{R}_{bb}$ are the input, the collateral, and the output matrices of dynamic stiffness, respectively. $\mathbf{R}, \mathbf{R}_{aa}$, $\mathbf{R}_{bb}$ are symmetric matrices.

$$\mathbf{R}_{ba} = \mathbf{R}_{ab}^{T}. \tag{5.6}$$

From Equations (5.1) and (5.4) we obtain

$$\begin{bmatrix} -\mathbf{Q}_a \\ \mathbf{Q}_b \end{bmatrix} = \begin{bmatrix} \mathbf{R}_{aa}\mathbf{R}_{ab} \\ \mathbf{R}_{ba}\mathbf{R}_{bb} \end{bmatrix} \begin{bmatrix} \mathbf{q}_a \\ \mathbf{q}_b \end{bmatrix}. \tag{5.7}$$

5.2.2 Mixed Dynamic Matrix

In some vibration problems, it is more convenient to use the so-called mixed dynamic matrix instead of the matrices of dynamic compliance (stiffness). Indeed, for example, in a system with symmetry, it is more convenient to divide the output vector of the displacement into two or more parts. Therefore, let us divide the *l-dimensional* input vector $\mathbf{q}_a$ into two vectors: l_1-dimensional vector $\mathbf{q}_{1a}$ and $(l - l_1)$-dimensional vector $\mathbf{q}_{2a}$. Let us also divide vector $\mathbf{q}_b$ into two vectors: t_1-dimensional vector $\mathbf{q}_{1b}$ and $(t-t_1)$-dimensional vector $\mathbf{q}_{2b}$. The two vectors of forces $\mathbf{Q}_a$ and $\mathbf{Q}_b$ are also divided in an analogous manner. Then the matrix Equation (5.1) has the form

$$\begin{bmatrix} \mathbf{q}_{1a} \\ \mathbf{q}_{2a} \\ \mathbf{q}_{1b} \\ \mathbf{q}_{2b} \end{bmatrix} = \begin{bmatrix} \mathbf{e}_{aa} & | & \mathbf{e}_{ab} \\ - & - & - \\ \mathbf{e}_{ba} & | & \mathbf{e}_{bb} \\ & | & \end{bmatrix} \begin{bmatrix} -\mathbf{Q}_{1a} \\ -\mathbf{Q}_{2a} \\ \mathbf{Q}_{1b} \\ \mathbf{Q}_{2b} \end{bmatrix}. \tag{5.8}$$

In this equation, each of the matrices $\mathbf{e}_{aa}$, $\mathbf{e}_{bb}$, $\mathbf{e}_{ab}$, $\mathbf{e}_{ba}$ is partitioned into four blocks. Permuting the rows and columns and solving this equation for vectors $\mathbf{q}_{2a}$, $\mathbf{q}_{2b}$, $\mathbf{Q}_{1a}$, and $\mathbf{Q}_{1b}$, we have

$$\mathbf{Y}_2 = \mathbf{P}\mathbf{Y}_1, \tag{5.9}$$

where

$$\mathbf{Y}_2 = \begin{bmatrix} \mathbf{q}_{2a} \\ \mathbf{q}_{2b} \\ -\mathbf{Q}_{1a} \\ \mathbf{Q}_{1b} \end{bmatrix}, \ \mathbf{Y}_1 = \begin{bmatrix} -\mathbf{Q}_{2a} \\ \mathbf{Q}_{2b} \\ \mathbf{q}_{1a} \\ \mathbf{q}_{1b} \end{bmatrix}, \tag{5.10}$$

$$\mathbf{P}= \begin{bmatrix} \mathbf{e}_{aa}\mathbf{e}_{ab} & \mathbf{S}_{aa}\mathbf{S}_{ab} \\ \mathbf{e}_{ba}\mathbf{e}_{bb} & \mathbf{S}_{ba}\mathbf{S}_{bb} \\ \tilde{\mathbf{S}}_{aa}\tilde{\mathbf{S}}_{ab} & \mathbf{R}_{aa}\mathbf{R}_{ab} \\ \tilde{\mathbf{S}}_{ba}\tilde{\mathbf{S}}_{bb} & \mathbf{R}_{ba}\mathbf{R}_{bb} \end{bmatrix}. \tag{5.11}$$

Or for short

$$\mathbf{P} = \begin{bmatrix} \mathbf{e} & \mathbf{S} \\ \tilde{\mathbf{S}} & \mathbf{R} \end{bmatrix}. \tag{5.12}$$

$\mathbf{P}$ is called a mixed dynamic matrix (MDM). Its diagonal blocks $\mathbf{e}$ and $\mathbf{R}$ are symmetric and $\tilde{\mathbf{S}} = \mathbf{S}^T$ due to the reciprocity theorem.

Let us now clarify the physical meaning of the obtained matrices. To this end, let us assume that in (5.9) and (5.10)

$$\mathbf{Q}_{2b} = \mathbf{q}_{1a} = \mathbf{q}_{1b} = 0.$$

Then we obtain

$$\mathbf{q}_{2a} = \mathbf{e}_{aa}\mathbf{Q}_{2b}. \tag{5.13}$$

It can be seen from this that $\mathbf{e}_{aa}$ is the input MDC under the condition that additional constraints have been imposed on this cell at which $\mathbf{q}_{1a} = \mathbf{q}_{1b} = 0$ (i.e., rigid connections at the input and output of the cell).

The matrices of collateral dynamic compliance $\mathbf{e}_{ab}, \mathbf{e}_{ba}$ and the output matrix of dynamic compliance $\mathbf{e}_{bb}$ are determined analogously. Therefore, matrix $\mathbf{e}$ (5.12) can be interpreted as a MDC of the cell with additionally imposed connections in the direction of all components of vectors $\mathbf{q}_{1a}$, $\mathbf{q}_{1b}$. The dynamic stiffness matrix $\mathbf{R}$ determines the dependence between the components of vectors $\mathbf{Q}_2$ and $\mathbf{q}_2$ under the condition that $\mathbf{Q}_1 = 0$. Matrix $\mathbf{S}$ determines the dependence between the components of vectors $\mathbf{q}_2$ and $\mathbf{q}_1$. We can treat it as a matrix of relative mechanical displacements.

Matrix $\tilde{\mathbf{S}}$ determines the dependence between the components of vectors $\mathbf{Q}_2$ and $\mathbf{Q}_1$. This matrix can be treated as a matrix of the relative dynamic forces.

Then matrix $\mathbf{P}$ (5.12) can be written in the form

$$\mathbf{P} = \begin{bmatrix} \text{MDC} & \mathbf{S} \\ \tilde{\mathbf{S}} & \text{MDS} \end{bmatrix}.$$

The elements of matrix $\mathbf{P}$ are either determined directly for the cell with additionally imposed connections or expressed by means of the matices $\mathbf{e}$ or $\mathbf{R}$ of the original cell.

5.2.3 Transition Matrix

This matrix can be considered as generalization of the quadrupole theory [63].

Let us consider the special case where the connectedness both at the input and the output of the cell is the same, i.e., $t = l = m$. Then all blocks of matrix $\mathbf{e}$ will be square matrices of the order $m \times m$. Solving Equation (5.1) for the output values $\mathbf{q}_b$, $\mathbf{Q}_b$, we obtain

$$\mathbf{X}_b = \mathbf{W}\mathbf{X}_a, \tag{5.14}$$

where

$$\mathbf{X}_a = \begin{bmatrix} \mathbf{q}_a \\ \mathbf{Q}_a \end{bmatrix}, \mathbf{X}_b = \begin{bmatrix} \mathbf{q}_b \\ \mathbf{Q}_b \end{bmatrix}, \tag{5.15}$$

$$\mathbf{W} = \begin{bmatrix} \mathbf{W}_{11} & \mathbf{W}_{12} \\ \mathbf{W}_{21} & \mathbf{W}_{22} \end{bmatrix}, \tag{5.16}$$

$\mathbf{X}_a$ and $\mathbf{X}_b$ are the vectors of the displacements and the forces at the input and output of the cell, and $\mathbf{W}$ is the transition matrix for the cell, its order is $(2m \times 2m)$.

From (5.1), (5.15), and (5.16) – under the condition that $l = t = m$ – the dependence between the blocks of the matrices $\mathbf{e}$ and $\mathbf{W}$ can be obtained.

$$\left.\begin{aligned}\mathbf{e}_{aa} &= \mathbf{W}_{21}^{-1}\mathbf{W}_{22},\\ \mathbf{e}_{ab} &= \mathbf{W}_{21}^{-1},\\ \mathbf{e}_{ba} &= \mathbf{W}_{11}\mathbf{W}_{21}^{-1}\mathbf{W}_{22} - \mathbf{W}_{12},\\ \mathbf{e}_{bb} &= \mathbf{W}_{11}\mathbf{W}_{21}^{-1},\end{aligned}\right\} \tag{5.17}$$

Analogously we can obtain

$$\begin{aligned}&\mathbf{W}_{11} = -\mathbf{R}^{-1}{}_{ab}\mathbf{R}_{aa}, \quad \mathbf{W}_{12} = -\mathbf{R}^{-1}{}_{ab}\\ &\mathbf{W}_{21} = \mathbf{R}_{ba} - \mathbf{R}_{bb}\mathbf{R}^{-1}{}_{ab}\mathbf{R}_{aa}, \quad \mathbf{W}_{22} = -\mathbf{R}_{bb}\mathbf{R}^{-1}{}_{ab}.\end{aligned} \tag{5.17a}$$

Since in $\mathbf{e}$ (and $\mathbf{R}$) the off-diagonal blocks are mutually transposed, so the blocks in transition matrix should also be not independent.

For example, it follows from

$$\mathbf{e}_{ba} = \mathbf{e}_{ab}^{T}$$

that

$$\mathbf{W}_{11}\mathbf{W}_{21}^{-1}\mathbf{W}_{22} - \mathbf{W}_{12} = \left[\mathbf{W}_{21}^{T}\right]^{-1}.$$

Let us now multiply the two sides of the equation by $\mathbf{W}_{21}^{T}$. Then we obtain

$$\mathbf{W}_{21}^{T}\mathbf{W}_{11}\mathbf{W}_{21}^{-1}\mathbf{W}_{22} - \mathbf{W}_{21}^{T}\mathbf{W}_{12} = \mathbf{E}, \tag{5.18}$$

where $\mathbf{E}$ is a identity square matrix ($m\times m$).

Equality (5.18) can be called a general condition of the quadrupole since at m=1 this condition turns into the equality

$$w_{11}w_{22} - w_{21}w_{12} = 1, \tag{5.19}$$

which is known in the quadrupole theory [63]

5.3 Finite Difference Equations

Let us consider two sequential s and $s+1$ cells of a system. Let us write the equilibrium condition and the equality of the displacements in the cross section between the cells as follows:

$$\mathbf{Q}_{bs} = \mathbf{Q}_{a,s+1} \tag{5.20}$$

$$\mathbf{q}_{a,s+1} = \mathbf{q}_{bs}. \tag{5.21}$$

Let us write Eq. (5.1) for the k cell in the new notations:

$$\begin{bmatrix} \mathbf{q}_{as} \\ \mathbf{q}_{bs} \end{bmatrix} = \begin{bmatrix} \mathbf{G}_s & \mathbf{F}^T{}_s \\ \mathbf{F}_s & \mathbf{T}_s \end{bmatrix} \begin{bmatrix} -\mathbf{Q}_{as} \\ \mathbf{Q}_{bs} \end{bmatrix}, \tag{5.22}$$

where

$$\mathbf{G}_s = \mathbf{e}_{aa}, \mathbf{F}^T{}_s = \mathbf{e}_{ab}, \mathbf{F}_s = \mathbf{e}_{ba}, \mathbf{T}_s = \mathbf{e}_{bb}.$$

From (5.21) and (5.22) for cells s and $s+1$ we have

$$-\mathbf{F}_s\mathbf{Q}_{as} + \mathbf{T}_s\mathbf{Q}_{bs} + \mathbf{G}_{s+1}\mathbf{Q}_{a,s+1} - \mathbf{F}^T{}_{s-1}\mathbf{Q}_{b,s+1} = 0. \tag{5.23}$$

For brevity's sake, we omit the subscript indices a and b at $\mathbf{Q}$ and after taking into account (5.20), we can write:

$$\begin{aligned} \mathbf{Q}_{as} &= \mathbf{Q}_{b,s-1} = \mathbf{Q}_{s-1} \\ \mathbf{Q}_{bs} &= \mathbf{Q}_{a,s+1} = \mathbf{Q}_s \qquad \text{etc.} \\ \mathbf{Q}_{a,s+2} &= \mathbf{Q}_{b,s+1} = \mathbf{Q}_{s+1}. \end{aligned}$$

Then Eq. (5.23) can be written as

$$-\mathbf{F}_s\mathbf{Q}_{s-1} + (\mathbf{G}_{s+1} + \mathbf{T}_s)\mathbf{Q}_s - \mathbf{F}_{s+1}\mathbf{Q}_{s+1} = 0. \tag{5.24}$$

Equation (5.24) is a canonic representation of the equation for dynamic compliances in matrix form. This equation can be considered also as a linear second-order matrix equation in finite differences.

With the new notations, the displacement vectors have the following form:

$$\left.\begin{aligned} \mathbf{q}_{s-1} &= \mathbf{G}_s(-\mathbf{Q}_{s-1}) + \mathbf{F}_s^T\,\mathbf{Q}_s, \\ \mathbf{q}_s &= \mathbf{F}_s(-\mathbf{Q}_{s-1}) + \mathbf{T}_s\mathbf{Q}_s = \mathbf{G}_{s+1}(-\mathbf{Q}_s) + \mathbf{F}^T{}_{s+1}\mathbf{Q}_{s+1}, \\ \mathbf{q}_{s+1} &= \mathbf{F}_{s+1}(-\mathbf{Q}_s) + \mathbf{T}_{s+1}\mathbf{Q}_{s+1}. \end{aligned}\right\} \tag{5.25}$$

In the case of forced vibrations, when the cells of the system are subjected to external exciting forces, Equation (5.24) takes the following form:

$$-\mathbf{F}_s\mathbf{Q}_{s-1} + (\mathbf{G}_{s+1} + \mathbf{T}_s)\mathbf{Q}_s - \mathbf{F}^T{}_{s+1}\mathbf{Q}_{s+1} + \mathbf{q}_s^f = 0, \tag{5.26}$$

where

$$\mathbf{q}_s^f = \mathbf{q}_{bs}^f - \mathbf{q}_{a,s+1}^f,$$

$\mathbf{q}_{bs}^f$ is the vector of the displacements at the output from cell s caused by the action of the exciting force and $\mathbf{q}_{a,s+1}^f$ is the analogous vector at the input to cell $s+1$.

Analogously, and using dynamic stiffness, we obtain a system of finite difference equations for displacements vector $\mathbf{q}_s$.

For free vibrations

$$\mathbf{R}_{bas}\,\mathbf{q}_{s-1} + [\mathbf{R}_{bbs} + \mathbf{R}_{aa,s+1}]\mathbf{q}_s + \mathbf{R}_{ab,s+1}\mathbf{q}_{s+1} = 0. \tag{5.27}$$

For forced vibrations

$$\mathbf{R}_{bas}\,\mathbf{q}_{s-1} + [\mathbf{R}_{bbs} + \mathbf{R}_{aa,s+1}]\mathbf{q}_s + \mathbf{R}_{ab,s+1}\mathbf{q}_{s+1} + \mathbf{Q}_s^f = 0, \tag{5.28}$$

where $\mathbf{R}_{b\alpha s}$, $\mathbf{R}_{bas}$, $\mathbf{R}_{aa,s+1}$, $\mathbf{R}_{ab,s+1}$ are the blocks of the fundamental matrix of dynamic stiffness

$$\mathbf{Q}_s^f = \mathbf{Q}_{bs}^f - \mathbf{Q}_{a,s+1}^f,$$

$\mathbf{Q}_{bs}^f$ is the vector of the reaction caused by the exciting force at the output from cell s , $\mathbf{Q}_{a,s+1}^f$ is the analogous vector at the input to cell $k+1$.

5.4 Mixed Dynamic Matrix as Finite Difference Equation

For the mixed dynamic matrix **P** (5.11), we obtain the following finite difference equation:

$$\mathbf{N}_{bas}\mathbf{Z}_{s-1} + [\mathbf{L}_{bbs} + \mathbf{L}_{aa,s+1}]\mathbf{Z}_s + \mathbf{N}_{ab,s+1}\mathbf{Z}_{s+1} = 0, \tag{5.28}$$

where

$$\left.\begin{aligned} \mathbf{N}_{bas} &= \begin{bmatrix} \tilde{\mathbf{e}}_{bas} & \mathbf{S}_{bas} \\ \mathbf{S}^T{}_{bas} & \mathbf{R}^T{}_{bas} \end{bmatrix}, & \mathbf{L}_{bbs} &= \begin{bmatrix} \tilde{\mathbf{e}}_{bbs} & \mathbf{S}_{bas} \\ \mathbf{S}_{bbs}^T & \mathbf{R}_{bas} \end{bmatrix} \\ \mathbf{L}_{aas} &= \begin{bmatrix} \tilde{\mathbf{e}}_{aas} & \mathbf{S}_{aas} \\ S_{aas}^T & \tilde{\mathbf{R}}_{aas} \end{bmatrix}, & \mathbf{N}_{abs} &= \begin{bmatrix} \tilde{\mathbf{e}}_{abs} & \mathbf{S}_{abs} \\ \mathbf{S}_{abs}^T & \tilde{\mathbf{R}}_{abs} \end{bmatrix}. \end{aligned}\right\} \tag{5.29}$$

$$\mathbf{N}_{abs} = \mathbf{N}_{bas}^T,$$

$$\mathbf{Z}_s = \begin{bmatrix} \mathbf{Q}_{1s} \\ \mathbf{q}_{2s} \end{bmatrix}, \ \mathbf{Q}_{1s} = \begin{bmatrix} -\mathbf{Q}_{1as} \\ \mathbf{Q}_{1bs} \end{bmatrix}, \ \mathbf{q}_{2s} = \begin{bmatrix} \mathbf{q}_{2as} \\ \mathbf{q}_{2bs} \end{bmatrix}. \tag{5.30}$$

The determination of all blocks and vectors that participate in Eqs. (5.28), (5.29), and (5.30) is presented above. The indices s–1, s, and $s+1$ indicate the number of the cell to which the matrix or vector belongs.

At forced vibrations, it is necessary to add to the left-hand side of Equation (5.30) terms that take into account the action of the exciting forces.

5.5 Transmission Matrix

We will obtain another pretty widely used type of system matrix – the transmission matrix. This matrix is a variant of the transition matrix discussed above. It connects the vectors of the forces and the displacements at a system's input and output. In accordance with Eqs. (5.14) and (5.15), it has the following form:

$$\mathbf{X}_n = \mathbf{W}^*\mathbf{X}_0 \,,$$

where $\mathbf{X}_0 = \begin{bmatrix} \mathbf{q}_0 \\ \mathbf{Q}_0 \end{bmatrix}$ is the input vector, $\mathbf{X}_n = \begin{bmatrix} \mathbf{q}_n \\ \mathbf{Q}_n \end{bmatrix}$ is the output vector,

$$\mathbf{W}^* = \begin{bmatrix} \mathbf{W}^*_{11} & \mathbf{W}^*_{12} \\ \mathbf{W}^*_{21} & \mathbf{W}^*_{22} \end{bmatrix}$$ are the blocks of matrix $\mathbf{W}^*$.

Let the system have n cells for which the transition matrices $\mathbf{W}_s{}^*$, $(s = 1 \ldots n)$ are known. Then the transmission matrix for the assembled system will be equal to the product of the transition matrices of all cells

$$\mathbf{W}^* = \mathbf{W}_n{}^*\mathbf{W}_{n-1}{}^* \ldots \mathbf{W}_2{}^*\mathbf{W}_1{}^*. \tag{5.31}$$

This method for obtaining the transmission matrix can be considered a matrix variant of the initial-parameter method. A large number of studies [20, 31] are devoted to this method. When the oscillation frequency has been preset and if the value of the displacements and the internal forces at the beginning of the section are known, this makes it possible to determine the same variables at the end of the section. In this way, by passing from section to section it is possible to reach the end of the system where the displacements and/or the internal forces are linked through the boundary conditions. And in this way, we obtain a characteristic equation and the equations for forced vibrations.

Chapter 6
Systems with Periodic Structure

6.1 Introduction

Let us consider several examples of mechanical systems with a periodic structure.

1. Periodic systems with a rectilinear axis. Each subsequent element of the system is obtained from the preceding one by a shift of some vector **a**. Systems of this type include a string with beads (Fig. 6.1a), a shaft with identical disks executing rotational or bending vibrations (Fig. 6.1b), a multispan beam with identical spans on rigid or elastic supports (Fig. 6.1c), a frame (Fig. 6.1d), a plane composite rod (Fig. 6.1e), a cylindrical shell with stiffening ribs and bulkheads, etc.
2. Periodic system with an axis with a constant curve radius.
   Each subsequent element of such a system is obtained from the preceding one by a rotation around the center at an angle α. A package of blades of an axial compressor or a turbine can serve as an example of such a system. This system is not closed (Fig. 6.2).
3. Periodic system with a rectilinear axis and a screwlike rotation. Each subsequent element of such a system is obtained from the preceding one by a shift of some vector **a** and a subsequent rotation at an angle α around an axis that is parallel to vector **a**. Certain structures of spatial crankshafts [56, 57] belong to periodic systems with a screwlike rotation. Figure 6.3 shows a three-throw crankshaft as an example of such a system.

The principal idea of periodic systems consists in the fact that it is possible to find the dynamic characteristics of the assembled system from the known dynamic characteristics of a single standard element.

L.Y. Banakh, M.L. Kempner, *Vibrations of Mechanical Systems with Regular Structure*, Foundations of Engineering Mechanics, DOI 10.1007/978-3-642-03126-7_6,

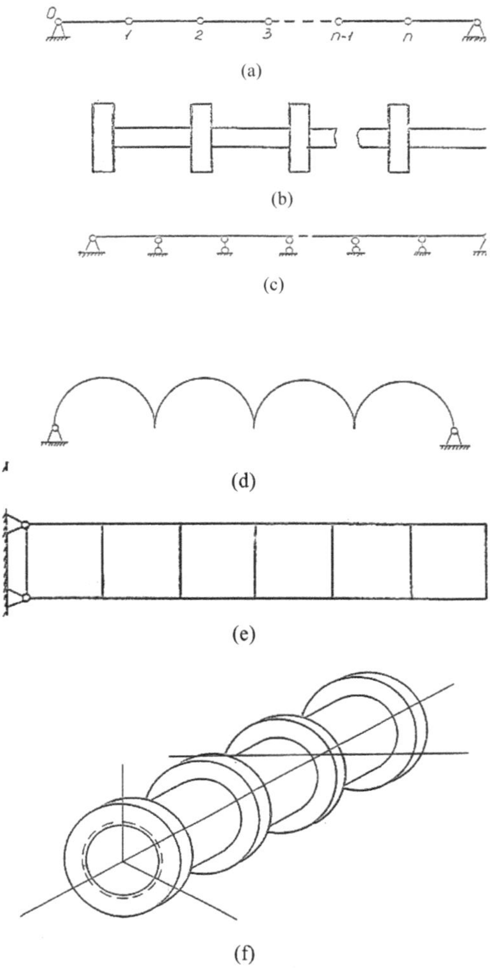

Fig. 6.1a-f Periodic systems with a rectilinear axis: string with beads (**a**), shaft with identical disks (**b**), multispan beam with identical spans (**c**), frame (**d**), plane composite rod (**e**), cylindrical shell with transversal stiffening ribs (**f**)

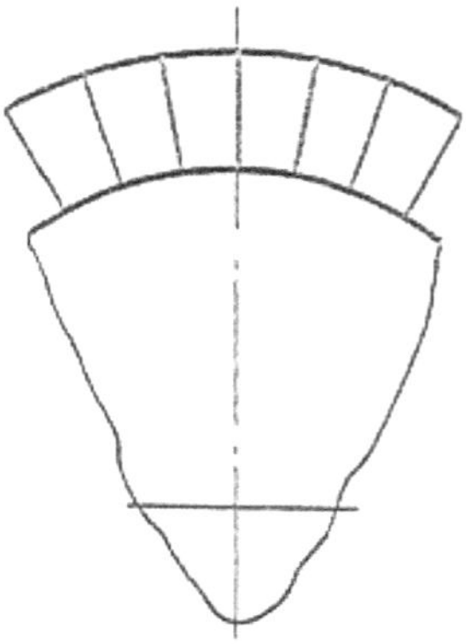

Fig. 6.2 Periodic system with an axis of constant curvature radius. A blades package

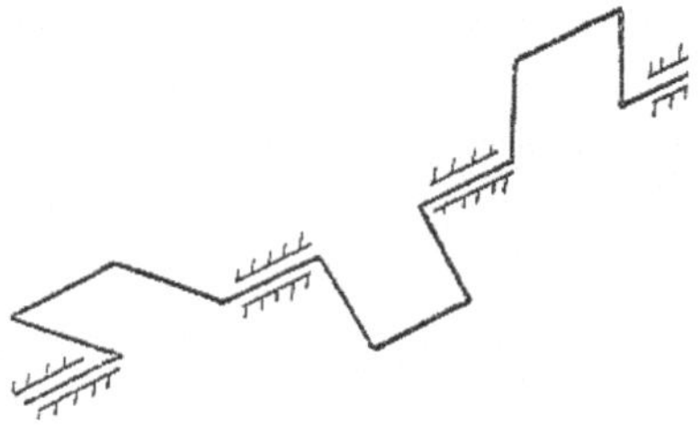

Fig. 6.3 Periodic system with a rectilinear axis and screwlike rotation. Spatial crankshaft

6.2 Dynamic Compliances and Stiffness for Systems with Periodic Structure

The finite-difference equation (5.24) for a not ramified (unidirectional) regular system has the following form:

$$-\mathbf{FQ}(s-1)+(\mathbf{G}+\mathbf{T})\mathbf{Q}(s)-\mathbf{\Phi Q}(s+1)=0, \qquad (6.1)$$

s is the number of the section. Due to the regularity of the system, the matrices **F**, **G**, **T**, and $\mathbf{\Phi}$ do not depend on the number s, $\mathbf{\Phi}=\mathbf{F}^T$. This equation is a homogeneous second-order finite-difference matrix equation with constant coefficients.

We will look for a particular solution of Eq. (6.1) in the following form (analogously to Chap. 3):

$$\mathbf{Q}(s)=\mathbf{AC}_sZ^s,\ Z=e^{\mu} \quad (s=0,1,2,\ldots..,n), \qquad (6.2)$$

where A is an m-dimensional vector and **C** an arbitrary constant vector.

From Eqs. (6.1) and (6.2) we obtain

$$[\mathbf{F}Z^{-1}-(\mathbf{G}+\mathbf{T})+\mathbf{\Phi}Z]\mathbf{A}=0. \qquad (6.3)$$

This equation can be considered a dispersion equation, analogous to that in Chap. 3. Indeed at use dynamic stiffness (compliances) beforehand it is supposed, that in system occur the harmonious vibrations $\mathbf{Q}=\mathbf{Q}_0e^{i\omega t}$. Therefore all coefficients in Eq. (6.3) depend on frequency ω which is considered as the current frequency. Therefore as against (2.2) here in Eq. (6.2) there is only the wave parameter μ determining length of a wave, which generally running wave. To solve the eigenvalue problem is necessary how usually to take into account boundary conditions.

Vector **A** is not equal to zero only if

$$\det(\mathbf{F}Z^{-1}-(\mathbf{G}+\mathbf{T})+\mathbf{\Phi}Z)=0. \qquad (6.4)$$

Equation (6.4) permits one to determine $2m$ characteristic values of Z_r ($r=1,2,\ldots\ldots,2m$). Let us consider all possible cases for characteristic values.

1. To begin with, let us assume that all roots Z_r are real and different. For each value ofZ_r, we will determine the vector with components $A_r^{(S)}$ (s = 1,2,. . ...., m) with an accuracy of up to a constant factor. We will write the general solution of Eq. (6.1) in the following form:

$$\begin{bmatrix} Q_1(s) \\ Q_2(s) \\ \cdot \\ \cdot \\ \cdot \\ Q_m(s) \end{bmatrix} = \begin{bmatrix} A_1^{(1)} \ldots . A_{2m}^{(1)} \\ A_1^{(2)} \ldots . A_{2m}^{(2)} \\ \ldots\ldots\ldots \\ \ldots\ldots\ldots \\ \ldots\ldots\ldots \\ A_1^{(m)} \ldots . A_{2m}^{(m)} \end{bmatrix} = \begin{bmatrix} C_1 Z_1^s \\ C_2 Z_2^s \\ \cdot \\ \cdot \\ \cdot \\ C_{2m} Z_{2m}^s \end{bmatrix}. \tag{6.5}$$

The $2m$ arbitrary constants C_r (r = 1,2,. . ..,2m) are determined from the boundary conditions.
2. Let us now assume that among the roots of Eq. (6.4) there are complex-conjugated pairs $Z_r = a_r + ib_r$, $Z_{r+1} = \bar{Z}_r$. These roots enter in Eq. (6.5) with complex-conjugated coefficients of the following form:

$$A_r^{(s)} C_r Z_r^k + A_{r+1}^{(S)} C_{r+1} \bar{Z}_r^k.$$

Therefore, if exist complex roots, the solution of Eq. (6.5) can be submitted as a linear combination of a function with real coefficients

$$A_r^{(s)} C_r Z_r^s + A_{r+1}^{(S)} C_{r+1} \bar{Z}_r^s = \mathrm{Re}(A_r^{(s)} C_r)\mathrm{Re}Z_r^s - \mathrm{Im}(A_r^{(s)} C_r)\mathrm{Im}Z_r^s. \tag{6.6}$$

3. Among the real roots of Eq. (6.4), there are also multiple ones.
Let us assume that the root Z_1 of Eq. (6.4) has a multiplicity of $l_1 > 1$. The solution of the homogeneous Eq. (6.1), which corresponds to a root with a multiplicity of l_1, can be represented as a linear combination of the solutions corresponding to these roots [42]

$$Q_s(s) = \sum_{r=0}^{l_1-1} A_r^{(s)} C_r Z_1^s \tag{6.7}$$

Let us pass now to determining the matrix of dynamic compliance for the assembled system. Let **G**, **T**, **F**, **Φ** be, respectively, the input, output, collateral transmission matrices of the system under consideration, and **q** (0) and **q**(n) the vector rows of displacement amplitudes at the input and output of the system, respectively.

Then

$$\begin{aligned} \mathbf{q}(0) &= -\mathbf{G}\mathbf{Q}(0) + \mathbf{\Phi}\mathbf{Q}(n) \\ \mathbf{q}(n) &= -\mathbf{F}\mathbf{Q}(0) - \mathbf{T}\mathbf{Q}(n). \end{aligned} \tag{6.8}$$

On the other hand:

$$\begin{aligned} \mathbf{q}(0) &= -\mathbf{G}\mathbf{Q}(0) + \mathbf{\Phi}\mathbf{Q}(1) \\ \mathbf{q}(n) &= -\mathbf{F}\mathbf{Q}(n-1) + \mathbf{T}\mathbf{Q}(n). \end{aligned} \tag{6.9}$$

The dynamic compliance matrices for the assembled system are determined from Eqs. (6.8) and (6.9) by taking into account (6.5), (6.6), and (6.7).

Let us introduce a new variable

$$\lambda = \frac{Z^{-1} + Z}{2} = \frac{e^{-\mu} + e^{\mu}}{2} \tag{6.10}$$

Then

$$Z_{1,2} = \lambda \pm \sqrt{\lambda^2 - 1}, \tag{6.11}$$

$$Z_{1,2}^{-1} = \lambda \mp \sqrt{\lambda^2 - 1}. \tag{6.12}$$

Equation (6.3) will take the form

$$[(\mathbf{F} + \mathbf{F}^T)\lambda + (\mathbf{F}^{\mathbf{T}} \text{-} \mathbf{F})\sqrt{1 - \lambda^2} - \mathbf{H}]\mathbf{A} = 0. \tag{6.13}$$

Therefore, the characteristic equation

$$\det[(\mathbf{F} + \mathbf{F}^T)\lambda + (\mathbf{F}^T - \mathbf{F})\sqrt{1 - \lambda^2} - \mathbf{H}] = 0, \tag{6.14}$$

where $\mathbf{F} + \mathbf{F}^T$ is a symmetric matrix, $\mathbf{F}^T - \mathbf{F}$ is a skew-symmetric matrix, and $\mathbf{H} = \mathbf{D} + \mathbf{T}$ is a symmetric matrix.

6.3 Dynamic Compliances of Single-Connectedness System

Let us consider now an important class of single-connectedness systems that are often used in various areas of engineering. For example, shaft executing the rotational or longitudinal vibrations, multispan beams, the blades of an axial compressor, and turbines with a free ring-type connection are single-connectedness systems. Quadrupole systems [63] are an analog of single-connectedness systems in the field of electrical engineering. Most systems in automatic control theory can also be considered as single-connectedness systems.

In order to determine the dynamic compliances of a single-connectedness system, we will use Eqs. (6.8),(6.9), and (6.13). Let *d, f, t* be the input, collateral, and output dynamic compliances of a regular element of a system consisting of *r* regular elements (connectedness m=1).

Taking into account that

$$f = f^T,$$

let us write Eq. (6.13) as a scalar one

$$2f\lambda - (d + t) = 0. \tag{6.15}$$

In the absence of damping, and depending on the frequency ω, the parameter λ can change in the range $-\infty < \lambda < \infty$, but it will always be a real number because *f, d,* and *t* are real values at any ω. Herein lies the difference between a single-connectedness system and a multiconnection one for which the roots λ, calculated from condition (6.14), can, generally speaking, be complex numbers.

Let us consider the following possible cases:

$$\begin{aligned}
&\text{a) } |\lambda| < 1,\\
&\text{b) } \lambda > 1,\\
&\text{c) } \lambda < -1,\\
&\text{d) } \lambda = 1,\\
&\text{e) } \lambda = -1.
\end{aligned}$$

Accordingly, using Eq. (6.10)

$$\begin{aligned}
&\text{a) } Z_{1,2} = \lambda \pm i\sqrt{1 - \lambda^2},\\
&\text{b) } Z_{1,2} = \lambda \pm i\sqrt{\lambda^2 - 1},\\
&\text{c) } Z_{1,2} = \lambda \pm i\sqrt{\lambda^2 - 1},\\
&\text{d) } Z_1 = Z_2 = 1,\\
&\text{e) } Z_1 = Z_2 = -1.
\end{aligned} \tag{6.16}$$

The general solution for *Q*(*s*) can be written for the cases under consideration in the following form:

$$\begin{aligned}
&\text{a) } Q(s) = C_1 \cos s\mu + C_2 \sin s\mu, \ \cos\mu = \lambda,\\
&\text{b) } Q(s) = C_1 chs\mu + C_2 shs\mu, \ ch\mu = \lambda,\\
&\text{c) } Q(s) = C_1(-1)^s chs\mu + C_2(-1)^s sh\, s\mu, \ ch\mu = -\lambda,\\
&\text{d) } Q(s) = C_1 + C_2 s\\
&\text{e) } Q(s) = C_1(-1)^s + C_2(-1)^s s,
\end{aligned} \tag{6.17}$$

where $\lambda = \lambda(\omega)$, $(s = 0, 1, \ldots, n)$,

Let us now consider two examples for determining the dynamic compliances of single-connectedness systems.

The arbitrary constants one may define as usually from boundary conditions. For example for case a) $|\lambda| < 1$ we receive

$$C_1 = Q(0),\ C_2 = \frac{Q(n)}{\sin\mu} - Q(0)\frac{\cos n\mu}{\sin\mu}.$$

Displacement in section s for single-connectedness system from Eq. (6.8) equals

$$q(s) = -dQ(s) + fQ(s+1).$$

Then we may receive the dynamic compliances for this system at $Q(n)=0$:

– input compliance $\tilde{d} = -\dfrac{q(0)}{Q(0)} = d - f\dfrac{\sin(n-1)\mu}{\sin n\mu}$

– output compliance $\tilde{t} = q(n)/Q(n) = t - f\dfrac{\sin(n-1)\mu}{\sin n\mu}$

– collateral compliance $\tilde{f} = q(0)/Q(n) = f\dfrac{\sin\mu}{\sin n\mu}$

Example 1. *Shaft executing the torsion vibrations. Let us isolate a regular subsystem (element) from the general system. This element represents a disk with moment of inertia θ and an inertia-free flexible section with rigidity* C. *The number of elements is equal to* n. *Let us calculate the dynamic compliances of this element.*

$$d = -\frac{1}{\theta\omega^2},\quad t = \frac{1}{C} - \frac{1}{\theta\omega^2},\quad f = -\frac{1}{\theta\omega^2},$$

where ω is the exciting frequency.

From Eq. (6.15) we obtain:

$$\lambda = 1 - \alpha,\qquad \alpha = \frac{\theta\omega^2}{2C}. \tag{6.18}$$

It can be seen that λ lies within the following ranges depending on parameter α:

$-1 < \lambda < 1$ at $0 < \alpha < 2$, $\lambda = -1$ at $\alpha = 2$, $\lambda = 1$ at $\alpha = 2$.

Therefore, three kinds of solutions are possible: a), c), and e). We will provide the result for the first case only:

$$\tilde{d} = -\frac{\cos(n\mu - \frac{\mu}{2})}{2C \sin\frac{\mu}{2} \sin n\mu}, \quad \tilde{t} = \frac{\cos(n\mu + \frac{\mu}{2})}{2C \sin\frac{\mu}{2} \sin n\mu},$$

$$\tilde{f} = \frac{\cos\frac{\mu}{2}}{2C \sin\frac{\mu}{2} \sin n\mu},$$

$$cos\, \mu = 1 - \alpha, \qquad 0 < \alpha < 2.$$

Example 2 *Multispan weightless beam with mass m in the middle of each span (Fig. 6.4a). Regular element – single-span beam with the mass in the middle (Fig. 6.4b). Dynamic compliances of element one can receive from Eqs. (1.15) [53],*

$$d = t = \frac{l}{3EJ}\frac{1-7\alpha}{1-16\alpha}, f = -\frac{l}{6EJ}\frac{1+2\alpha}{1-16\alpha},$$

$$\alpha = \frac{ml^3\omega^2}{768EJ},$$

l is the length of the span, EJ is the bending stiffness of the beam, and ω is the exciting frequency.

From Eq. (6.15) we obtain:

$$\lambda = -2\frac{1-7\alpha}{1+2\alpha}$$

Figure 6.4c shows a graph of the function $\lambda=\lambda(\alpha)$:

At $2<\alpha<\frac{1}{16}\lambda, \quad \lambda>1$ solution “a”

At $\alpha = \frac{1}{16}, \quad \lambda = 1$ solution “b”

At $\frac{1}{16}<\alpha<\frac{1}{4}, \quad |\lambda|<1$ solution “c”

At $\alpha = \frac{1}{4}, \quad \lambda = -1$ solution “d”

At $\alpha = \frac{1}{4}, \quad \lambda = 1$ solution “e”

For case “a” we have

$$d = t = -\frac{l}{6EJ}\frac{1+2\alpha}{1-16\alpha} ctgn\, \mu \sin\mu,$$

$$f = -\frac{l}{6EJ}\frac{1+2\alpha}{1-16\alpha}\frac{\sin\mu}{\sin n\mu},$$

$$\cos\mu = \lambda.$$

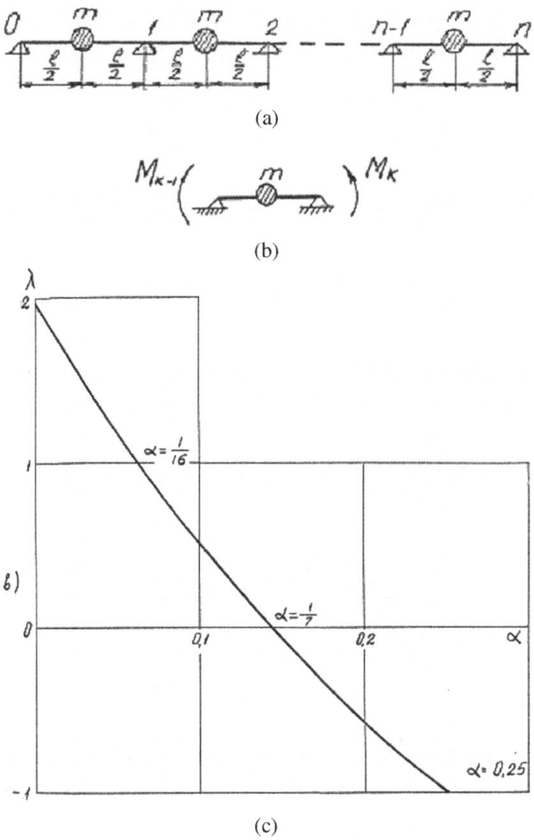

Fig. 6.4 Determining the dynamic compliance of a multispan weightless beam with concentrated masses

6.4 Transition Matrix

Let us now consider the solution of the above-discussed problem obtained with the help of a transition matrix.

Let us write Equality (4.14) for the element with number k

$$\mathbf{X}(s) = \mathbf{W}_s\mathbf{X}(s-1)\ (\mathrm{s} = 0, 1, 2, \ldots.., n), \tag{6.19}$$

where

$$\mathbf{X}(s) = \begin{bmatrix} \mathbf{q}(s) \\ \mathbf{Q}(s) \end{bmatrix},$$

$\mathbf{X}(s)$ is a $2m$-dimensional vector, $\mathbf{q}$ (s) and $\mathbf{Q}$ (s) are m-dimensional vectors of the displacements and the forces, respectively.

For a regular system, the transition matrix $\mathbf{W}_s = \mathbf{W}$ does not depend on the number of the element; that is why Equality (6.19) for the whole system will have the form:

$$\mathbf{X}(n) = \mathbf{W}^n \mathbf{X}(0) \tag{6.19a}$$

Instead of the successive multiplication of matrices, the following approach can be used [18]. Let us find the characteristic values λ of matrix $\mathbf{W}$ from the equation

$$\det(\mathbf{W} - \lambda \mathbf{E}) = 0. \tag{6.20}$$

Equation (6.20) has $2m$ real or complex-conjugated roots. Let all roots be different. Then a normal vector $\mathbf{U}^{(r)}$ with components $U_s^{(r)}(s = 1, 2, 3, \ldots\ldots, 2m)$ determined with an accuracy of up to an arbitrary factor will correspond to each characteristic value $\lambda_r (r = 1, 2, \ldots\ldots, 2m)$. Then

$$\mathbf{W} = \mathbf{U}^{-1} \mathbf{\Lambda} \mathbf{U}, \ \mathbf{\Lambda} = \operatorname{diag}[\lambda_r], \ \mathbf{U} = [\mathbf{U}^{(1)} \, \mathbf{U}^{(2)} \, _{\ldots} \mathbf{U}^{(2m)}].$$

Here the columns of matrix $\mathbf{U}$ (modal columns) represent the normal vectors $\mathbf{U}^{(r)}$.
For the nth degree of matrix $\mathbf{W}$, we will write the following equality:

$$\mathbf{W}^n = (\mathbf{U}^{-1} \mathbf{\Lambda} \mathbf{U})(\mathbf{U}^{-1} \mathbf{\Lambda} \mathbf{U}) \ldots (\mathbf{U}^{-1} \mathbf{\Lambda} \mathbf{U}) = \mathbf{U}^{-1} \mathbf{\Lambda}^n \mathbf{U}. \tag{6.21}$$

This relationship offers significant advantages if $m \ll n$.
To determine the nth degree of matrix $\mathbf{W}$, it is also possible to use Sylvester's theorem [18]

$$\mathbf{W}^n = \sum_{r=1}^{n} \lambda_r \mathbf{Z}_r, \tag{6.22}$$

where

$$Z_r = \frac{\prod_s (\mathbf{W} - \lambda_s \mathbf{E})}{\prod (\lambda_r - \lambda_s)}, \ (s = 1, 2, \ldots \ \ldots, r-1, r+1, \ldots\ldots, 2m). \tag{6.23}$$

If Eq. (6.20) has multiple roots, then in the general case matrix $\mathbf{W}$ cannot be reduced to a diagonal form. However, Jordan's form [41] can be used for obtaining $\mathbf{W}^n$ by using Jordan's diagonal blocks (cells).

6.5 Forced Vibrations

When calculating the forced vibrations for the sth element of the regular system, we have

$$\mathbf{X}(s) = \mathbf{W}\mathbf{X}(s-1) + \mathbf{\Psi}(s). \tag{6.24}$$

Here $\mathbf{\Psi}(s)$ is a $2m$-dimensional vector taking into account the external harmonic forces applied to an element with number s. This vector can include not only force-related but also kinematic excitation. Values of $\mathbf{\Psi}(s)$ for external harmonic excitation in the case of a rod working in conditions of extension-contraction, rotation, and bending are provided in [20, 47].

The general solution for inhomogeneous systems of difference equations (6.24) can be obtained by applying the method of variation of arbitrary constants. This solution has the following form:

$$\mathbf{X}(s) = \mathbf{W}(s)\mathbf{X}(0) + \sum_{j=1}^{s} \mathbf{W}(s-j)Y(j), \tag{6.25}$$

where $\mathbf{W}(s)$ is the fundamental matrix of a homogeneous system (6.24)

$$\mathbf{W}(s) = \mathbf{U}\begin{bmatrix} \lambda_1^s & & & \\ & \lambda_2^s & & \\ & & \ddots & \\ & & & \lambda_{2m}^s \end{bmatrix}\mathbf{U}^{-1} \tag{6.26}$$

λ_r – $(r = 1, 2, \ldots.., 2m)$ are the characteristic values of the transition matrix $\mathbf{W}$, $\mathbf{U}$ is the matrix of the normal vectors (6.22).

6.6 Vibrations of Blades Package

There are a number of structures where the blades of a gas turbine or an axial compressor are connected to each other. Frequently, this connection is of a ring type. However, in some structures, in particular in structures of stationary turbine engineering, groups of blades are integrated in packages. We will assume in the sequel that in all cases the linking belt and the blades form a unitary frame structure. The conditions for linking the shroud and the blades can change, but there should be no sliding between the shroud and the blades as well as between the individual components of the shroud.

The structure in which all blades are connected through a ring link is a system with cyclic symmetry. Such systems are discussed in the next chapter. Here we look at a case where a group of blades is connected in a package and the system is periodic. The regular element of the system will consist of a blade and a portion of the connection belt. The transition matrix $\mathbf{W}$ of such an element is determined from the expression

$$\mathbf{W} = \mathbf{W_e}\mathbf{W_b}, \tag{6.27}$$

where $\mathbf{W_b}$ is the transition matrix of the connection (shroud) and $\mathbf{W_e}$ the transition matrix of the blade.

The transition matrix of the connection $\mathbf{W_b}$ is determined by taking into account its elastic and inertia properties. Usually, only the elastic properties are taken into account.

In order to determine the natural frequencies of the package consisting of n blades (Fig. 6.2), we use the results obtained above. Let us add at the right-hand side of the last blade a section of the shroud that does not change the operating conditions of the structure if the inertia properties of the shroud are not taken into account. Let us write in block form Eq. (6.19a)

$$\begin{bmatrix} \mathbf{q}(n) \\ \mathbf{Q}(n) \end{bmatrix} = \begin{bmatrix} \mathbf{W}_{11}^n & \mathbf{W}_{12}^n \\ \mathbf{W}_{21}^n & \mathbf{W}_{22}^n \end{bmatrix} \begin{bmatrix} \mathbf{q}(0) \\ \mathbf{Q}(0) \end{bmatrix},$$

where $\mathbf{q}$ (0) and $\boldsymbol{q(n)}$ are the displacement vectors, and $\mathbf{Q}$ (0) and $\mathbf{Q}$ $(\boldsymbol{n})$ are the forces vectors.

Using the boundary conditions

$$\mathbf{Q}(n) = \mathbf{Q}(0) = 0$$

we obtain

$$\mathbf{W}_{21}^n\, \mathbf{q}(0) = 0.$$

Therefore, the frequency equation

$$\det\left(\mathbf{W}_{21}^n\right) = 0. \tag{6.28}$$

The elements of matrix $\mathbf{W}_{21}^n$ are determined with the help of Eq. (6.21). It is also possible to use (4.17).

$$\mathbf{W}_{21} = e_{a\alpha}^{-1}.$$

Let us consider a package where the blades are connected by means of an one ring link. It is assumed here that the connection has only extension-contraction mode. Such a package belongs to single-connectedness systems.

The regular element of the system is a blade with a section of a ring link. The frequency of its free vibrations is determined from the Eq. (6.17a)

$$\sin n\mu = 0, \tag{6.29}$$

where

$$\mu = \arccos \frac{d+t}{2f}.$$

This relationship corresponds to a zero value of the dynamic stiffness of the package. Here $d = e$ is the input dynamic compliance of the element, and $t = e + \frac{1}{c}$ is the output dynamic compliance of the element.

After a transformation

$$\mu = \arccos\left(1 + \frac{1}{2ce}\right), \tag{6.30}$$

e is the dynamic compliance of the blade and c is the static stiffness of the connecting section at extension-contraction.

6.7 Collective Vibrations of Blades

When simulating the vibrations of a blades package, the shroud is usually assumed to be elastic and inertia free and its weight is added to the weight of the blades. The shroud is subjected to bending deformations only. Figure 6.6 shows the sin-phased tangential vibrations of a package (one to three vibration forms) consisting of ten blades [109]. Analogous vibration modes of individual blades correspond to each sin-phased mode. Figure 6.7 shows torsion-axial vibrations of the blades. In addition to the tangential vibrations, torsion vibrations of the package are also possible (Fig. 6.8).

It can be seen from Figs. 6.6–6.8 that there exists some interaction between the vibration modes of the blades and the shroud, which represents a superposition of waves (as described in Chap. 3). Here also the determining factor is the ratio between the rigidity of the blades and the shroud in some mode of vibration. Therefore, the tangential vibrations of the blades emerge earlier than the corresponding vibrations of the shroud as its rigidity is greater than the rigidity of the blades in these modes of vibrations. At the same time, the shroud, on the contrary, has a lower rigidity to torsion than the blades. Then its torsion-axial vibrations arise earlier. All lowest modes are shown in Fig. 6.7. A similar picture occurs also at torsion vibrations (Fig. 6.8).

The bending – torsion vibrations of blades package on Fig. 6.9 are shown [65].

The first mode corresponds to axial in phase blades vibrations in plane of maximal stiffness + transversal shroud vibrations. The second mode: bending vibrations of blades + torsion vibrations of shroud. The third mode relates to bending vibrations of shroud with two nodes; the fourth – with three nodes. At the fifth mode there are blades vibrations with one node and torsion vibrations of shroud.

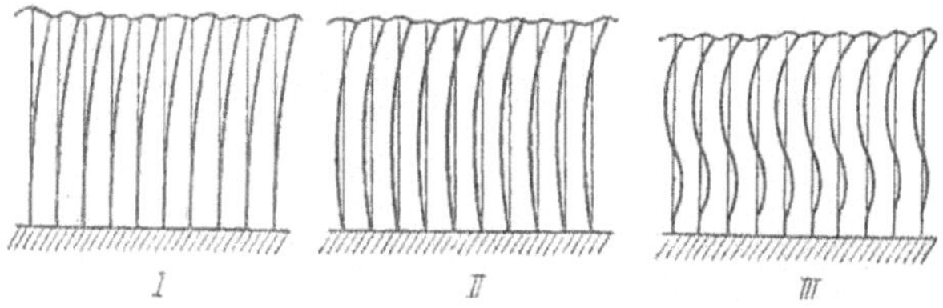

Fig. 6.6 Sin-phased tangential vibrations of blades

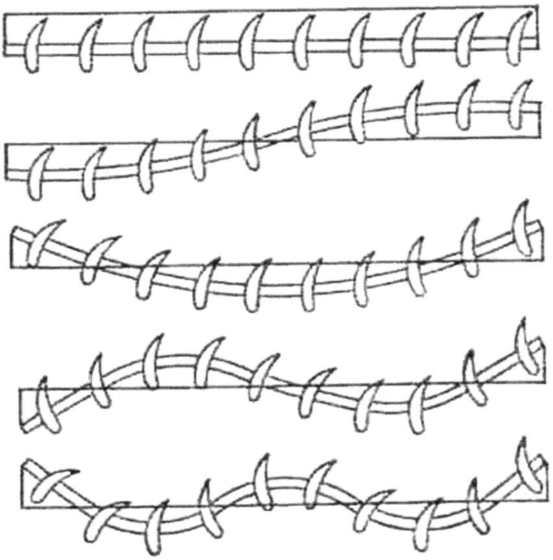

Fig. 6.7 Torsion-axial vibrations of the blades and the shroud

Fig. 6.8 Torsion vibrations of blade package

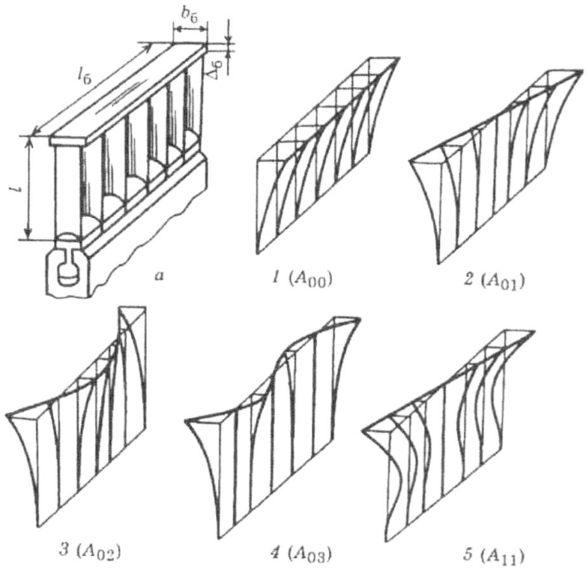

Fig. 6.9 Torsion-bending vibrations of blades package

Chapter 7
Systems with Cyclic Symmetry

7.1 Natural Frequencies and Normal Modes for Systems with Cyclic Symmetry

Some examples of systems with cyclic symmetry that are widely used in the mechanical engineering are shown in Fig. 7.1. They are a set of blades with a ring connection (Fig. 7.1a), a lifting propeller of a helicopter (Fig. 7.1b), radio telescopes, domes, towers, etc.

Such systems are studied from the position of the general theory of periodic systems. We will consider a system with cyclic symmetry as a special case of a generalized not ramified periodic system. In this case, we can use the transition matrices. For a continual system, it is necessary to replace the transition matrix with the *displacement operator.*

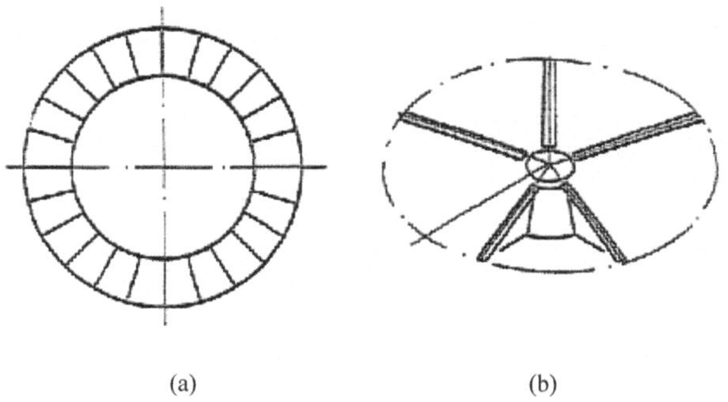

Fig. 7.1a, b Systems with cyclic symmetry

7.1.1 Natural Frequencies

Let a system with cyclic symmetry consist of n elements. The number of connections between the elements is equal to m. Let us write the equality for the element number s.

L.Y. Banakh, M.L. Kempner, *Vibrations of Mechanical Systems with Regular Structure*, Foundations of Engineering Mechanics, DOI 10.1007/978-3-642-03126-7_7,

$$\mathbf{X}(s) = \mathbf{W}\mathbf{X}(s-1) \quad (s = 1, 2, \ldots, n), \tag{7.1}$$

where $\mathbf{W}$ is a transition matrix that does not depend on the number s.

According to Chap. 5, for vector $\mathbf{X}(0)$ we have

$$\mathbf{X}(n) = \mathbf{W}^n \mathbf{X}(0). \tag{7.2}$$

For systems with cyclic symmetry, the output of the nth element coincides with the input of the first element, i.e.,

$$\mathbf{X}(n) = \mathbf{X}(0). \tag{7.3}$$

Substituting (7.3) in (7.2) we obtain

$$\left(\mathbf{W}^n - \mathbf{E}\right)\mathbf{X}_0 = 0.$$

The condition for a nontrivial solution is the equality

$$\det(\mathbf{W}^n - \mathbf{E}) = 0. \tag{7.4}$$

This means that the equality $\mathbf{W}^n = \mathbf{E}$ corresponds to the free vibrations of the system.

Let $\lambda_1, \lambda_2, \ldots \lambda_{2m}$ be the characteristic values of matrix $\mathbf{W}$. In this case:

$$\det(\mathbf{W} - \lambda\mathbf{E}) = (\lambda - \lambda_1)(\lambda - \lambda_2)\ldots(\lambda - \lambda_{2m}). \tag{7.5}$$

Then $\lambda_1^n, \lambda_2^n, \ldots, \lambda_{2m}^n$ will be the characteristic values of matrix $\mathbf{W}^n$ [41] and

$$\det\left(\mathbf{W}^m - \mu\mathbf{E}\right) = (\mu - \mu_1)(\mu - \mu_2)\ldots(\mu - \mu_{2m}), \tag{7.6}$$

where

$$\mu_r = \lambda_r^n \qquad (r = 1, 2\ldots, 2m).$$

Comparing (7.4) and (7.6), we see that $\mu = 1$; then

$$\det(\mathbf{W}^m - \mathbf{E}) = (1 - \mu_1)(1 - \mu_2)\ldots(1 - \mu_{2m}). \tag{7.7}$$

Equality (7.4) is satisfied if at least one of the characteristic values of matrix $\mathbf{W}^n$ is equal to one

$$\mu_r = 1 \qquad (r = 1, 2, \ldots, 2m)\,.$$

However, since $\mu_r = \lambda_r^n$, then

$$\lambda_r = \sqrt[n]{1} = e^{i\frac{2\pi}{n}r}\,(r = 1, 2, \ldots, n)\,,$$

or

$$\lambda_r = \cos\frac{2\pi}{n}r + i\sin\frac{2\pi}{n}r. \tag{7.8}$$

At even values of n, the two characteristic values at $r = n \quad (\lambda_n = 1)$ and $r = \frac{n}{2} \quad \left(\lambda_{\frac{n}{2}} = -1\right)$ are real. The complex characteristics values λ_{n-r} and λ_r are complex-conjugated

$$\lambda_{n-r} = \overline{\lambda_r}\left(1 \le r \le \frac{n}{2}\right). \tag{7.9}$$

In this way, the characteristic values can be determined, instead of Eq. (7.4), from the solution of the equation

$$\det\left[(\mathbf{W} - \lambda_r E)\right] = 0 \tag{7.10}$$

for all values of λ_r, calculated from Eq. (7.8).

We transform expression (7.10) into a form that is more convenient for calculation of the characteristic values. If Equality (7.10) is satisfied, then the following equality will also be satisfied:

$$\det(\mathbf{W} - \overline{\lambda_r}\mathbf{E}) = 0.$$

Then

$$\det\left[(\mathbf{W} - \lambda_r\mathbf{E})\left(\mathbf{W} - \overline{\lambda_r}\mathbf{E}\right)\right] = 0,$$

or

$$\det\left(\mathbf{W}^2 - 2\mathbf{W}\cos\frac{2\pi}{n}r + \mathbf{E}\right) = 0. \tag{7.10a}$$

Replacing $\cos\frac{2\pi}{n}r$ with $1 - 2\sin^2\frac{\pi r}{n}$ we obtain

$$\det\left[(\mathbf{W} - \mathbf{E})^2 + 4\mathbf{W}\sin^2\frac{\pi r}{n}\right] = 0. \tag{7.11}$$

The elements of matrixff $\mathbf{W}$ depends from ω. Therefore the values of ω that satisfy condition (7.4) are the frequencies of the free vibrations of the system. In other words, all values of ω that turn into unit at least one of the characteristic values of matrix $\mathbf{W}^n$, namely $\mu_r = \lambda_r^n = 1 (r = 1 \ldots n)$ are frequencies of the free vibrations of the system.

7.1.2 Normal Modes

Let us now proceed to determine the natural modes of vibrations. A vector $\mathbf{X_r}(0)$ will correspond in the input of the first element to each value of λ_r. This vector is determined with accuracy up to a constant coefficient from the equation

$$(\mathbf{W} - \lambda_r \mathbf{E})\,\mathbf{X}_r(0) = 0. \tag{7.11a}$$

Any linear combination of vectors $\mathbf{X}_r(0)$ and $\mathbf{X}_{n-r}(0)$ determines the normal mode corresponding to the obtained value of ω. For vector $\mathbf{X}_r(k)$, we will have

$$\begin{aligned}
&\mathbf{X}_r(k) = [a\mathrm{Re}\,(\mathbf{X}_r(0)) - bJm\,(\mathbf{X}_r(0))]\cos ks + \\
&+ [aJm\,(\mathbf{X}_r(0)) + b\mathrm{Re}\,(\mathbf{X}_r(0))]\sin ks
\end{aligned},$$

$$k = 0, 1, 2, \ldots, n-1; \quad s = \frac{2\pi}{n} r,$$

$$r = 0, 1, \ldots, \frac{n}{2} \text{ at an even } n,$$

$$r = 0, 1, \ldots, \frac{n-1}{2} \text{ at an odd } n,$$

a and b are arbitrary scalar factors which one define from boundary conditions.

7.2 Vibrations of Blades System

7.2.1 Different Designs of Blades Connecting

Gas turbines and axial compressors usually use shroud structures where the blades are connected to each other by means of a ring connection. A blade and shroud element is shown in Figs. 7.2a, b and 7.3a, b.

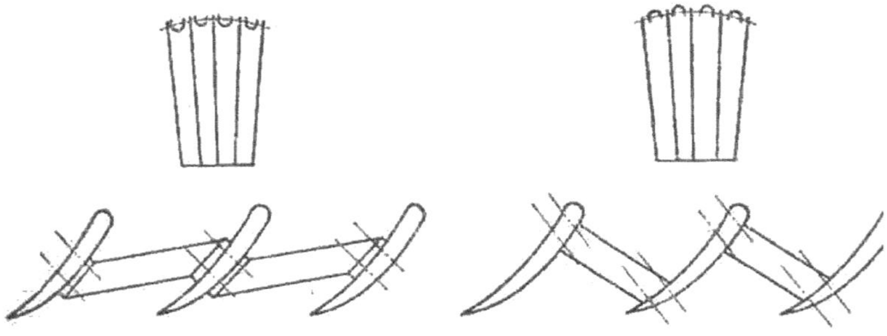

(a) (b)

Fig. 7.2a, b Different designs of connecting blades to a shroud

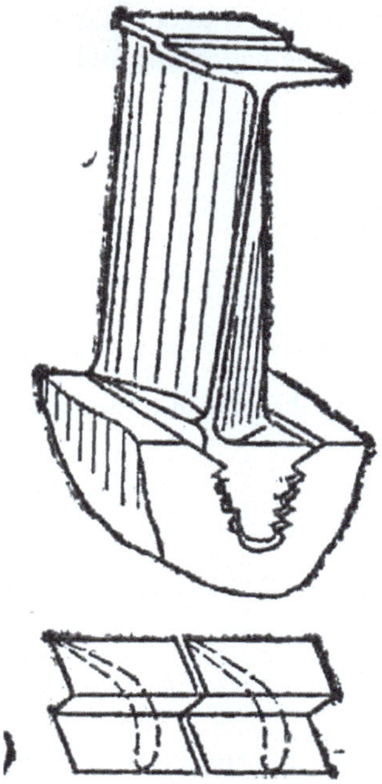

Fig. 7.3 Connecting elements using friction forces

Blades are connected to each other with the help of rod elements. Only the conditions of linking the shroud and the blades can change, but there should be no sliding between the shroud and the blades as well as between the individual components of the connection (Fig. 7.2a, b). When other designs of connecting the elements are used, their interaction is ensured by the dry friction forces, which emerge under the effect of the centrifugal forces or by the tightness in the mounting (Fig. 7.3).

However, in all cases, the connection and the blades form a unitary frame structure. In this way, this structure is a system with cyclic symmetry.

When studying the vibrations of blades with a ring connection in some structures, it is more convenient to use a frequency equation expressed by its dynamic compliances (stiffness).

7.2.2 Natural Frequencies for Blades System

Let us write Eq. (7.11) by replacing the dynamic stiffness with dynamic compliances:

$$\det\left[\mathbf{e}_{ba}\mathbf{Z}^{-1} + (\mathbf{e}_{ba} + \mathbf{e}_{aa}) + \mathbf{e}_{ab}\mathbf{Z}\right] = 0,\ \mathbf{e}_{ab} = \mathbf{e}_{ba}. \tag{7.12}$$

Assuming

$$\mathbf{Z} = \mathbf{A}e^{\lambda s} \tag{7.13}$$

and taking into consideration that from the conditions of ring periodicity λ must be an imaginary value multiple 2π, we obtain

$$\lambda = 2s\pi i,$$

then

$$\det\left[2\mathbf{e}_{ab}\cos\frac{2s\pi}{n} - (\mathbf{e}_{ab} + \mathbf{e}_{aa})\right] = 0.$$

In the case of a single-connectedness system when $\mathbf{e}_{aa}, \mathbf{e}_{ab}$ are scalar we obtain

$$2e_{ab}\cos\frac{2s\pi}{n} - (e_{ab} + e_{aa}) = 0. \tag{7.14}$$

For a blade with a free ring connection

$$e_{ab} = e_{aa} = e, \quad e_{aa} = \frac{1}{c} + e,$$

where c is the stiffness of the ring connection element ($s = 1, 2, \ldots, \frac{n}{2}$).
From Eq. (7.14) we obtain:

$$4\sin^2\frac{s\pi}{n} = -\frac{1}{ec} = -\frac{R}{c} \tag{7.14a}$$

where R is the dynamic stiffness of the blade.

7.2.3 Normal Modes for Blades System

Taking into consideration the relationships (7.13) and (7.14), we see that the normal modes of the blades have the shape shown in Fig. 7.4 [109].

1. The displacements and angles of rotation are discrete harmonic functions of the view $A_i \sin(m\varphi_i + \alpha_i)$, where m is the number of nodal diameters, φ_i the angular coordinate, and A_i, φ_i the amplitude and phase shifts, respectively.
2. Out-of-phase vibrations with a different number of nodal diameters m on a circle.

The normal modes are divided into two groups. The first group includes the normal modes in which the blade itself has one node of vibration (in fixed point). The second group includes vibrations in which the blade has two nodes of vibration, etc. The case $m = 0$ corresponds to cophased vibrations. In the case of an odd number of blades, the nth frequency of the cophased mode is not multiple. In the case of an even number of blades, there is one additional cophased mode corresponding to

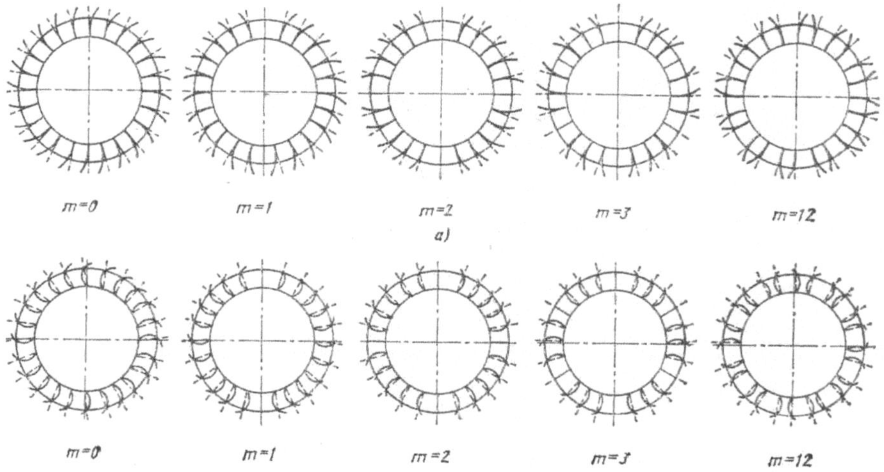

Fig. 7.4 Vibration modes of blades

$m = n/2$. At this mode, all neighboring blades vibrate in antiphase (this corresponds to the results of Chap. 4).

7.3 Numerical and Experimental Results for Blades with Shroud

Let us now consider several examples for determining the vibrations frequency for blades with different designs of fixing them to a shroud.

7.3.1 Free Ring Connection

Blades with a ring connection are shown in Fig. 7.5a. Let us assume that the ring connection is located in the end section of the blades and each blade has been replaced by a single-mass system.

Let the number of blades be $n = 12$ with the relative compliance of the blade set at $\frac{3EJ}{l^3C} = 0.5$.

It is necessary to determine the frequencies of the simultaneous vibrations and angular velocity of a shaft under resonance in the field of centrifugal forces.

(a) The frequencies of the simultaneous vibrations in the absence of a centrifugal force field are obtained from (7.14a)

$$4\sin^2\frac{s\pi}{n} = -\frac{3EJ}{l^3C}\left(1-\frac{\omega^2}{\omega_0^2}\right), \omega_0^2 = \frac{3EJ}{l^3C}. \quad (7.15)$$

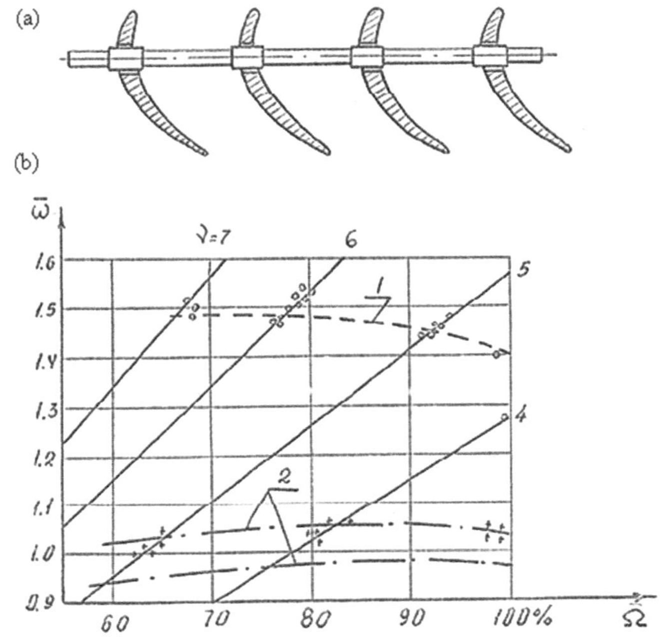

Fig. 7.5a, b Numerical and experimental results for blades with a ring connection: 1 – with a shroud; 2 – without a shroud

Solving Eq. (7.15) at $s = 1, 2, \ldots, 6$ and introducing the dimensionless frequency $\bar{\omega} = \frac{\omega}{\omega_0}$, we obtain the following values for the frequencies of the simultaneous vibrations:

$$\bar{\omega}_1 = 1.24;\ \bar{\omega}_2 = 1.73;\ \bar{\omega}_3 = 2.23;\ \bar{\omega}_4 = 2.64;\ \bar{\omega}_5 = 2.9;\ \bar{\omega}_6 = 3.0.$$

As is known, vibration modes are orthogonal to each other. Then the exiting force with frequency, for example, equal to the second natural frequency can cause blades vibrations on this frequency only. In Fig. 7.6 we have on the abscissa the dimensionless frequency $\bar{\Omega} = \frac{\Omega}{\omega_0}$, where Ω is the angular velocity of shaft rotation; on the ordinate, the dimensionless frequency $\bar{\omega} = \frac{\omega}{\omega_0}$. The rays of excitation from $\nu = \omega/\Omega$, $\nu = 1 \ldots 6$ are built on dimensionless coordinates $\bar{\Omega} = \frac{\Omega}{\omega_0}, \bar{\omega} = \frac{\omega}{\omega_0}$. These rays have points deposited on them whose ordinates are equal to the relative frequencies of the vibrations with the same number as the exciting rays (the bottom curve in Fig. 7.6). So, for example, the ordinate point of a ray with number 4 is equal to $\bar{\omega}_4 = 2.64$. The curve connecting these points is drawn with a dotted line.

(b) Let us now see how the angular velocity changes at a resonance by taking into consideration the centrifugal forces of the blades.

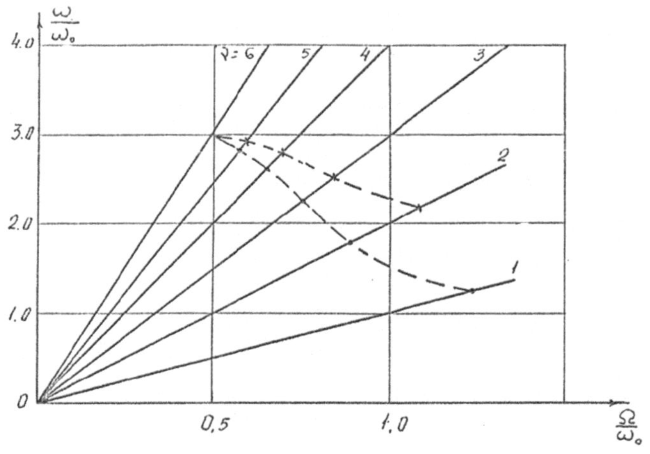

Fig. 7.6 Simultaneous vibrations of blades

In this case, the frequency equation looks like

$$4\sin^2\frac{s\pi}{n}=\frac{B}{C},$$

$$4\sin^2\frac{s\pi}{n}=-\frac{3EJ}{l^3C}\bar{\Omega}^2\left[\frac{1}{\frac{l}{r_n}\left(1-\frac{th\bar{\Omega}\sqrt{3\frac{r_n}{l}}}{\bar{\Omega}\sqrt{3\frac{r_n}{l}}}\right)}-1-\left(\frac{\bar{\omega}}{\bar{\Omega}}\right)^2\right] \tag{7.16}$$

where $s=1,2,\ldots,6$.

For the parameters provided above, we find the values of $\bar{\Omega}$ at $\nu=\frac{\omega}{\Omega}=1,2,\ldots,6$.

The values of $\bar{\Omega}$ corresponding to the resonance frequencies from each of the excitation harmonics are plotted in Fig. 7.6 (the points of this curve are marked with crosses -x-x-x-x-). It is positioned above that absence of centrifugal forces leads to an increase in the total stiffness of the system. It should be pointed out that for the first mode ($\nu=1$) there is no resonance, which in Fig. 7.6 means the absence of points on the ray $\nu=1$. As is known, such a case is observed when there are vibrations in the centrifugal force field of an isolated blade.

Figure 7.5 shows the numerical and experimental results for blades with a ring shrouding positioned in the plane of rotation. These are the frequencies of the simultaneous vibrations of blades connected by a ring shroud (Fig. 7.5). The frequencies were determined from Eq. (7.16) using the experimentally obtained dynamic stiffness (Chap. 2).

7.3.2 Blades with Paired-Ring Shroud

An element of a system is shown in Fig. 7.7a.

(a)

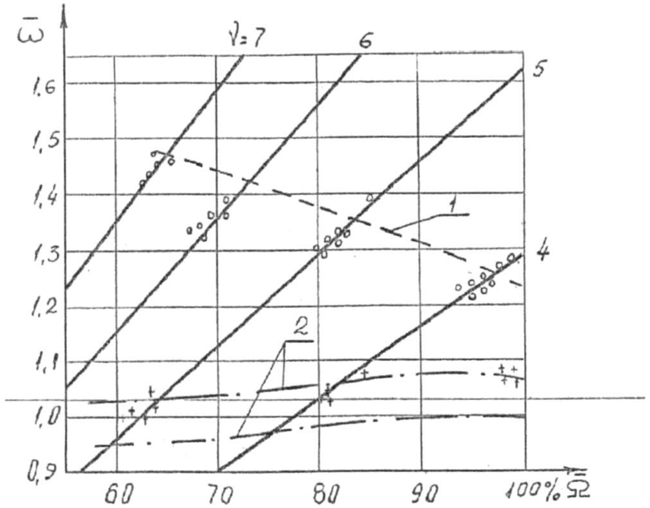

(b)

Fig. 7.7a, b Blades with paired-ring connection (**a**), experimental and numerical results for the natural frequencies (**b**)

From Eq. (7.14) we obtain

$$4\sin^2\frac{s\pi}{n} = -\left[\frac{1}{ke_{ab}} + \frac{e_{aa}}{e_{ab}} + \frac{e_{bb}}{e_{ab}}\right],$$

where e_{aa} is the input dynamic compliance, e_{ab} the collateral one, e_{bb} the output compliance, and k the stiffness of the shroud to stretching-compression. Figure 7.7b shows the experimental and numerical results for the natural frequencies. The determination of the natural frequencies was performed both for an absolutely rigid shroud and for the case where its compliance is taken into account. Analyzing the obtained results, we can draw the conclusion that the frequencies in the case of a paired-ring connection are lower than in the regular case. This means that in this case the effect of the torsion compliance of the blade sharply increases. Therefore, it plays a dominant role, while at the same time the compliance of the shroud already does not affect, in practical terms, the frequency of the simultaneous vibrations.

7.3.3 Blades Shrouded by Shelves

This method of banding is shown in Fig. 7.8. The shelves are implemented as a single unit together with the blade. At vibrations, these shelves can be treated as a compact inextensible ring performing bending vibrations in its plane.

We will consider two types of simultaneous vibrations of blades and a connecting ring.

The first type of vibration is characterized by tangential displacements of the gravity centers of the blade sections at locations where they are connected to a shelf. Due to the absolute stiffness of the ring, all blade displacements are the same. Therefore, the study of simultaneous vibrations of blades can be replaced by studying the vibrations of an isolated blade but with an additional elastic connection at the location where it is connected to the shelf. This connection only impedes the rotation of the section around the z axis.

In the second type of sections vibrations on which a shroud has been mounted, there are no tangential displacements. In this way, the blades and the shroud form ring system of two-connectedness. The following frequency equation is valid for such a system:

$$\det\left[(\mathbf{W}-\mathbf{E})^2 + 4\sin^2\frac{s\pi}{n}\mathbf{W}\right] = 0, \quad (s = 1, 2 \ldots n/2),$$

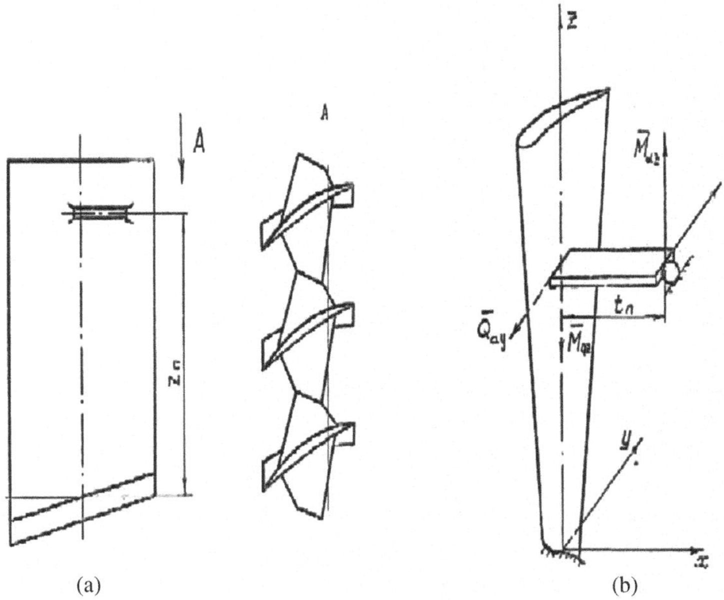

Fig. 7.8 a, b Blades banded with shelves (**a**) and a regular blade element (**b**)

where

$$\mathbf{W} = \mathbf{W}_{sh}\mathbf{W}_b,$$

$\mathbf{W}_{sh}$ is the transition matrix of the shroud, and $\mathbf{W}_b$ is the transition matrix of the blade.

At this point the following assumptions are introduced:

$$v(z_n) = 0,\ Q_z = M_x = M_y = 0,$$

where z_n is the distance between the base of the blade and the point of its connection to the shelf.

Let us now write the transition matrix at the input to and the output from the blade, i.e., in the section where the shroud is fastened:

$$\begin{bmatrix} u_b \\ \theta_b \\ Q_{by} \\ M_{bz} \end{bmatrix} = \begin{bmatrix} 1 & W_{21} & e_{aa} & e_{ba} \\ W_{12} & 1 & e_{ab} & e_{bb} \\ R_{aa} & R_{ba} & 1 & 0 \\ R_{ab} & R_{bb} & 0 & 1 \end{bmatrix} \begin{bmatrix} u_a \\ \theta_a \\ Q_{ay} \\ M_{az} \end{bmatrix}$$

where $e_{ij}, R_{ij} = e_{ij}^{-1}$ are, respectively, the dynamic compliance and stiffness of the blade,

$$M_{ba} = -(e_{aa}R_{ba} + e_{ba}R_{bb}), M_{ab} = -(e_{ab}R_{aa} + e_{bb}R_{ab}).$$

The transition matrix for the section of the shroud is calculated as for weightless rod with a constant cross section:

$$\mathbf{W}_{sh} = \begin{bmatrix} 1 & l & \frac{t_n^2}{2E'J_{sh}} & \frac{t_n^3}{6E'J_{sh}} \\ 0 & 1 & -\frac{t_n}{E'J_{sh}} & -\frac{t_n^2}{2E'J_{sh}} \\ 0 & 0 & 1 & -l \\ 0 & 0 & 0 & 1 \end{bmatrix}$$

where t_n is the length of the shroud between the blades (the step of the shroud)and J_b is the bending stiffness of the shroud.

Figure 7.9 shows the numerical and the experimental results for blades of a compressor

We have on the abscissa the relative rotation frequency $\bar{\Omega} = \frac{\Omega}{\Omega_{\max}}$, where Ω is the angular speed of the shaft rotation. On the ordinate $\bar{\omega} = \frac{\omega}{\omega_0}$, where ω_0 is the "static" frequency of the blade. It is determined in the absence of support on the shelf. The resonance frequencies with different harmonics of excitation $\nu = 2, 4, \ldots, 16$ are connected, by convention, by a flat curve. The bottom

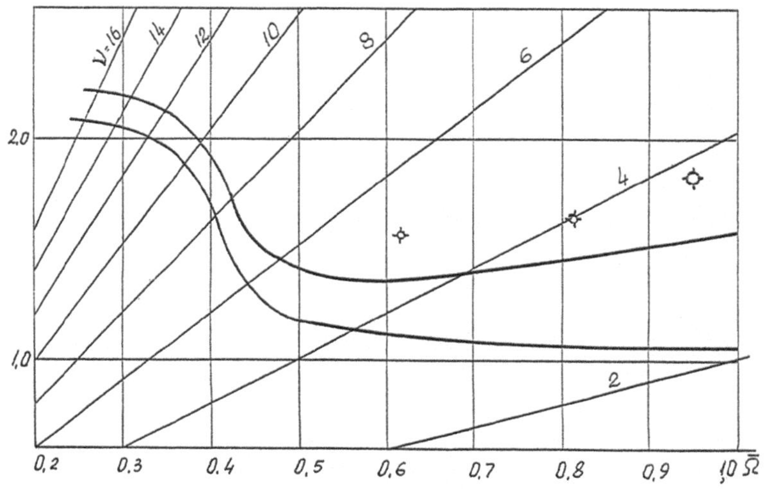

Fig. 7.9 Numerical and experimental frequencies for the compressor blades with shelves (0 – experimental points)

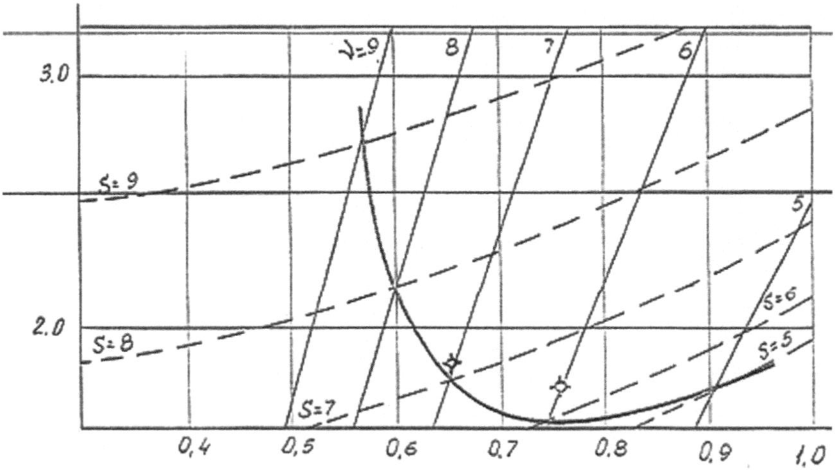

Fig. 7.10 Numerical and experimental natural frequencies of the turbine blades a with shelves (0 – experimental point)

calculation curve is constructed without taking into account the centrifugal force field, and the top one by taking this effect into account. The experimental points are positioned slightly higher than the top theoretical curve.

Figure 7.10 shows analogous results for the blade of a turbine. The bottom curve is absent. Curves showing the dependence of the frequency on the angular velocity with different numbers of waves $s = 5, 6 \ldots, 9$ (dotted lines) have been added. The other notations are the same as in Fig. 7.9.

Some more designs of shroud one can find in [68].

Chapter 8
Systems with Reflection Symmetry Elements

8.1 Reflection Symmetry Element and Its Dynamic Characteristics

Let us consider a chain system of n regular elements (cells). Let each of the system elements have a symmetry plane Π-Π (Figs. 8.1 and 8.2). The symmetry is understood both in the geometrical and the physicomechanical sense of the word. All connections of an element at the input into and the output from it must be symmetric with respect to Π-Π. We will call the system that satisfies the above-mentioned conditions an unidirectional (chain) system with reflection symmetry elements. A homogeneous elastic body with a constant cross section (string, beam, cylindrical shell) with some additional regular point-type "inclusions," for example concentrated masses, rigid and elastic supports, elastic internal connections, etc., can be considered as a special case of such systems.

Let the reflection symmetry element (Fig. 8.1) have normal modes that are symmetric with respect to plane Π-Π. In this case, part of the generalized coordinates included in input vector $\mathbf{q}_a$ coincides in value and sign with the corresponding generalized coordinates included in output vector $\mathbf{q}_b$. The remaining coordinates are also identical with each other but have opposite signs. Let us agree to relate all coordinates with the same sign in one group and those with opposite signs in a second group. Let us designate the vector that includes the coordinates of the first group as vector $\mathbf{q}_{1a}(\mathbf{q}_{1b})$ and the one for the coordinates of the second group as vector $\mathbf{q}_{2a}(\mathbf{q}_{2b})$. Let us differentiate into two groups the components of the internal forces

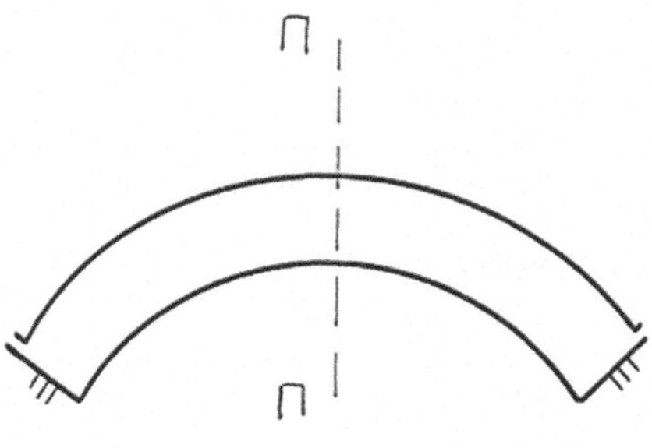

Fig. 8.1 Reflection symmetric

L.Y. Banakh, M.L. Kempner, *Vibrations of Mechanical Systems with Regular Structure*, Foundations of Engineering Mechanics, DOI 10.1007/978-3-642-03126-7_8,

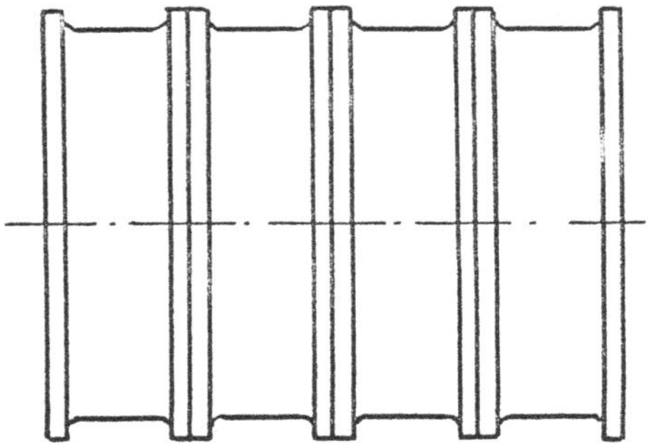

Fig. 8.2 Compressor case model element

using the same property; the sign rule for these, however, will be the opposite of the sign rule used for the coordinates (see Chap. 5). Therefore, the first group will include the component vectors $\mathbf{Q}_{2a}$ and $\mathbf{Q}_{2b}$, and the second group $\mathbf{Q}_{1a}$ and $\mathbf{Q}_{1b}$, respectively.

For normal modes that are skew-symmetric with respect to plane Π, the picture will be just the opposite.

As an example of a reflection symmetry element, let us consider a rod executing bending-longitudinal vibrations in its plane (Fig. 8.3). The first group includes the vertical displacements of the gravity centers of the end sections V_a and V_b, the bending moments M_a, M_b, the normal forces N_a, N_b; the second group includes the longitudinal displacements U_a and U_b, the rotation angles ϕ_a and ϕ_b, and the transversal forces Q_a and Q_b[1].

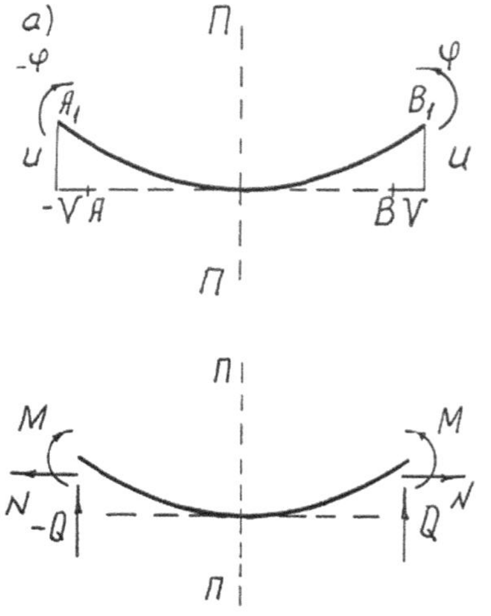

(a)

(b)

Fig. 8.3 Displacements (**a**) and internal forces (**b**) at the ends of the symmetric vibrations of a rod

[1] It should be noted, that at use of a part 1 methods (including FEM) there is no necessity to divide the variables on two groups. It turns out automatically by appropriate choice of coordinates axes direction.

8.1.1 Dynamic Stiffness and Compliance Matrices for Reflection Symmetry Element

Let us write a relation between the vectors of the displacements and the forces using the dynamic stiffness matrix.

$$\begin{bmatrix} -\mathbf{Q}_{1a} \\ -\mathbf{Q}_{2a} \\ \mathbf{Q}_{1b} \\ \mathbf{Q}_{2b} \end{bmatrix} = \mathbf{R} \begin{bmatrix} \mathbf{q}_{1a} \\ \mathbf{q}_{2a} \\ \mathbf{q}_{1b} \\ \mathbf{q}_{2b} \end{bmatrix}. \tag{8.1}$$

Thereby

$$\begin{bmatrix} \mathbf{q}_{1b} \\ \mathbf{q}_{2b} \end{bmatrix} = -\mathbf{I} \begin{bmatrix} \mathbf{q}_{1a} \\ \mathbf{q}_{2a} \end{bmatrix}, \begin{bmatrix} \mathbf{Q}_{1b} \\ \mathbf{Q}_{2b} \end{bmatrix} = \mathbf{I} \begin{bmatrix} \mathbf{Q}_{1a} \\ \mathbf{Q}_{2a} \end{bmatrix}, \mathbf{I} = \begin{bmatrix} -\mathbf{E} & \\ & \mathbf{E} \end{bmatrix}. \tag{8.2}$$

The stiffness matrix has the view

$$\mathbf{R} = \begin{bmatrix} \mathbf{R}_{aa} & \mathbf{R}_{ab} \\ \mathbf{R}_{ba} & \mathbf{R}_{bb} \end{bmatrix} = \begin{bmatrix} \mathbf{R}_{11} & -\mathbf{R}_{12} & \mathbf{B}_{11} & \mathbf{B}_{12} \\ -\mathbf{R}_{21} & \mathbf{R}_{22} & \mathbf{B}_{21} & \mathbf{B}_{22} \\ \mathbf{B}_{11} & -\mathbf{B}_{12} & \mathbf{R}_{11} & \mathbf{R}_{12} \\ -\mathbf{B}_{21} & \mathbf{K}_{22} & \mathbf{R}_{21} & \mathbf{R}_{22} \end{bmatrix}. \tag{8.3}$$

The dynamic stiffness matrix for a reflection symmetry element has some specific characteristics compared to matrix (5.2) for an ordinary element of a regular system. The diagonal blocks of this matrix coincide with an accuracy up to the sign with elements $\mathbf{R}_{12}$ and $\mathbf{R}_{21}$. Blocks $\mathbf{B}_{11}$, $\mathbf{B}_{22}$ are symmetric. Blocks $\mathbf{B}_{12}$, $\mathbf{B}_{21}$ are mutually transposed and have opposite signs, $\mathbf{B}_{12} = -\mathbf{B}_{21}^T$.

It should be noted that every deformed state of the reflection symmetry element can be represented by superposing displacements that are symmetric and skew-symmetric (antisymmetric) with respect to plane Π. Each of the elements of matrix $\mathbf{R}$ can be presented in the form of a half-sum or a half-difference of two summands. The first summand is a reaction in the connection caused by a unitary symmetric displacement of the node; the second summand is a reaction in the connection caused by an analogous skew-symmetric displacement. Correspondingly, we will use the notations '$\underset{\cdot}{-}$' and '$\underset{\cdot}{=}$' for their superscript indices.

$$\mathbf{R}_{bb} = \frac{\bar{\mathbf{R}}_{bb} + \bar{\bar{\mathbf{R}}}_{bb}}{2}, \tag{8.4}$$

$$\mathbf{R}_{aa} = \mathbf{I}\mathbf{R}_{bb}\mathbf{I}, \tag{8.5}$$

$$\mathbf{R}_{ab} = \frac{\mathbf{I}}{2}\left[\bar{\mathbf{R}}_{bb} - \bar{\bar{\mathbf{R}}}_{bb}\right], \tag{8.6}$$

$$\mathbf{R}_{ba} = \frac{1}{2}\left[\bar{\mathbf{R}}_{bb} - \bar{\bar{\mathbf{R}}}_{bb}\right]\mathbf{I}, \tag{8.7}$$

where

$$\mathbf{I} = \begin{bmatrix} -\mathbf{E} & 0 \\ 0 & \mathbf{E} \end{bmatrix}. \tag{8.8}$$

Analogous relations are valid also for the dynamic compliance matrix, which is the inverse of the dynamic stiffness matrix:

$$\mathbf{R}{=}\mathbf{e}^{-1}, \begin{bmatrix} \mathbf{q}_{1a} \\ \mathbf{q}_{2a} \\ \mathbf{q}_{1b} \\ \mathbf{q}_{2b} \end{bmatrix} = \mathbf{e} \begin{bmatrix} -\mathbf{Q}_{1a} \\ -\mathbf{Q}_{2a} \\ \mathbf{Q}_{1b} \\ \mathbf{Q}_{2b} \end{bmatrix} = \begin{bmatrix} \mathbf{G}_{11} & -\mathbf{G}_{12} & \boldsymbol{\Phi}_{11} & \boldsymbol{\Phi}_{12} \\ -\mathbf{G}_{21} & \mathbf{D}_{22} & \boldsymbol{\Phi}_{21} & \boldsymbol{\Phi}_{11} \\ \boldsymbol{\Phi}_{11} & -\boldsymbol{\Phi}_{12} & \mathbf{G}_{11} & \mathbf{G}_{12} \\ -\boldsymbol{\Phi}_{21} & \boldsymbol{\Phi}_{22} & \mathbf{G}_{21} & \mathbf{G}_{22} \end{bmatrix} \begin{bmatrix} -\mathbf{Q}_{1a} \\ -\mathbf{Q}_{2a} \\ \mathbf{Q}_{1b} \\ \mathbf{Q}_{2b} \end{bmatrix}.$$

8.1.2 Mixed Matrix for Reflection Symmetry Element

Let us now determine the mixed matrix **P** (5.11) connecting the two vectors $\mathbf{Y}_1$ and $\mathbf{Y}_2$:

$$\mathbf{Y}_2 = \mathbf{P}\mathbf{Y}_1.$$

We will include in vector $\mathbf{Y}_1$ all coordinates of the displacements and the internal forces of the first group and in vector $\mathbf{Y}_2$ those of the second group.

$$\mathbf{Y}_1 = \left[\mathbf{Q}_{2a}\ \mathbf{Q}_{2b}\ \mathbf{q}_{1a}\ \mathbf{q}_{1b}\right]^T,$$
$$\mathbf{Y}_2 = \left[\mathbf{q}_{2a}\ \mathbf{q}_{2b}\ -\mathbf{Q}_{1a}\ \mathbf{Q}_{1b}\right]^T.$$

Using the special structure of matrix **R** of the reflection symmetry element and inverting the blocks of this matrix, we obtain the following relation for matrix **P**:

$$\mathbf{P} = \begin{bmatrix} \mathbf{G} & \mathbf{G}^* & \mathbf{S} & \mathbf{S}^* \\ \mathbf{G}^* & \mathbf{G} & \mathbf{S}^* & -\mathbf{S} \\ -\mathbf{S}^T & -\mathbf{S}^{*T} & \mathbf{B} & \mathbf{B}^* \\ \mathbf{S}^{*T} & \mathbf{S}^T & \mathbf{B}^* & \mathbf{B} \end{bmatrix} \tag{8.9}$$

where

$$\begin{aligned}
&\mathbf{G} = \left(\mathbf{R}_{22} - \mathbf{B}_{22}\mathbf{R}_{22}^{-1}\mathbf{B}_{22}\right)^{-1}, \\
&\mathbf{G}^* = \left(\mathbf{B}_{22} - \mathbf{R}_{22}\mathbf{B}_{22}^{-1}\mathbf{R}_{22}\right)^{-1}, \\
&\mathbf{S} = \mathbf{R}_{12}\mathbf{G} - \mathbf{B}_{12}\mathbf{G}^*, \\
&\mathbf{S}^* = \mathbf{R}_{12}\mathbf{G}^* - \mathbf{B}_{12}\mathbf{G}, \\
&\mathbf{B} = \mathbf{R}_{11} - \mathbf{R}_{12}\mathbf{S}^{\mathrm{T}} + \mathbf{R}_{12}\mathbf{S}^{*\mathrm{T}}, \\
&\mathbf{B}^* = \mathbf{B}_{11} - \mathbf{B}_{12}\mathbf{S}^{*\mathrm{T}} - \mathbf{B}_{12}\mathbf{S}^{\mathrm{T}}.
\end{aligned} \tag{8.9a}$$

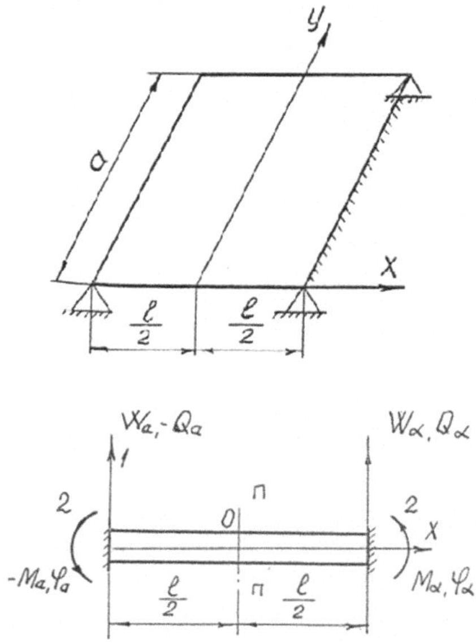

Fig. 8.4 Determining the dynamic stiffness of a rectangular plate

It follows from Eqs. (8.9) and (8.9a) that matrices **G**, **G***, **B**, and **B*** are symmetric. This property of the matrix will makes it possible to obtain later "symmetrized" equations (i.e., equations for the coordinates describing symmetric normal modes) in finite differences.

Let us consider a numerical example.

Let us have a rectangular plate with constant thickness (Fig. 8.4a,b). Its two opposite sides $y = 0, y = a$ are hinge-supported, while the other two $x = \pm l/2$ are rigidly fixed.

The plate is a reflection symmetry element with respect to plane Π. Let us apply to one of the plate edges an exciting force according to the harmonic law $w\left(\frac{l}{2}, y\right) = 1 \cdot \sin \frac{\pi n y}{a}$ $(n = 1, 2, 3, \ldots)$. Then deflection w, being a function of the coordinate y, will now be a continual, not a discrete, variable. Let us calculate the reactions arising at the edges of the plate that will represent its dynamic stiffness.

The expression for the amplitude of bending during vibrations can be presented in the following form:

$$w\,(x, y) = (C_1 ch\alpha_1 x + C_2 sh\alpha_1 x + C_3 sh\alpha_2 x + C_4 sh\alpha_2 x) \sin \frac{\pi n y}{a}, \qquad (8.10)$$

where

$$\alpha_{1,2}^2 = \alpha^2 \pm \frac{n^2 \pi^2}{a^2} (n = 1, 2, 3 \ldots), \qquad (8.11)$$

By this

$$\alpha^2 - \frac{\pi^2 n^2}{a^2} > 0.$$

If $\alpha^2 - \dfrac{\pi^2 n^2}{a^2} < 0$, then solution (8.10) should be taken in the form:

$$w\,(x, y) = (C_1 ch\,|\alpha_1|\,x + C_2 sh\,|\alpha_1|\,x + C_3 \cos \alpha_2 x + C_4 \sin \alpha_2 x) \sin \frac{\pi n y}{a}, \quad (8.10a)$$

$$\alpha^4 = \omega^2 \frac{\rho h}{D} \quad (8.12)$$

$D = \dfrac{E' h^3}{12\left(1 - \mu^2\right)}$ is the cylindrical stiffness, E' is the elasticity modulus of the first type, μ is the Poisson coefficient, h is the plate thickness, ρ is the density of the material, ω is the vibration frequency.

Later we will also need the rotation angle θ of the normal to the middle surface of the plate with respect to the y axis

$$\theta = \frac{\partial w}{\partial x} \quad (8.13)$$

For deformations that are symmetric with respect to plane Π

$$\bar{w}\ (x) = C_1 ch\alpha_1 x + C_3 ch\alpha_2 x, \quad (8.14)$$

$$\bar{Q}\,(x) = C_1 \alpha_1 sh\alpha_1 x - C_3 \alpha_2 \sin \alpha_2 x. \quad (8.15)$$

For skew-symmetric ones

$$\bar{\bar{w}}\ (x) = C_2 ch\alpha_1 x + C_4 ch\alpha_2 x, \quad (8.16)$$

$$\bar{\bar{Q}}\ (x) = C_2 \alpha_1 ch\alpha_1 x + C_4 \alpha_2 \cos \alpha_2 x. \quad (8.17)$$

$$\bar{w} + \bar{\bar{w}} = w, \quad \bar{Q} + \bar{\bar{Q}} = Q.$$

We determine the bending moment and the transversal force from the following equations by observing the sign rule:

$$\left.\begin{aligned} M &= D\left(\frac{\partial^2 w}{\partial x^2} + \mu \frac{\partial^2 w}{\partial y^2}\right) \\ Q &= -D\left[\frac{\partial^3 w}{\partial x^3} + (2 - \mu)\,\frac{\partial^3 w}{\partial x \partial y^2}\right] \end{aligned}\right\} \quad (8.18)$$

Taking into account that $w\,(x, y) = w\,(x) \sin \dfrac{\pi n}{a} y$, we obtain

$$\left.\begin{aligned} M &= D\left[\frac{\partial^2 w(x)}{\partial x^2} - \mu\frac{n^2\pi^2}{a^2}w(x)\right]\sin\frac{n\pi}{a}y \\ Q &= -D\left[\frac{\partial^3 w(x)}{\partial x^3} - (2-\mu)\frac{n^2\pi^2}{a^2}\frac{dw(x)}{dx}\right]\sin\frac{n\pi y}{a} \end{aligned}\right\}. \tag{8.19}$$

Let us assign a number to each coordinate and force, as is done in Fig. 8.4б: w and Q – number (1), φ and M– number (2). In order to obtain the dynamic stiffness at the right-hand edge (the output) of the plate, we have to determine the arbitrary constants in expressions (8.14)–(8.17) and then introduce them in (8.19) and set $x = \frac{l}{\lambda}$. Then we find

$$\bar{R}_{11} = -\frac{\alpha^2\alpha_1\alpha_2}{D}\frac{\sin\alpha^*{}_1\sin\alpha^*{}_2}{L_1},$$

$$\bar{R}_{12} = \bar{R}_{21} = \frac{(1-\mu)\,\pi^2\cdot n^2}{2\,Da^2} + \alpha^2\frac{\alpha_2 h\alpha^*{}_1\sin\alpha^*{}_2 - \alpha_1 sh\alpha^*{}_1\cos\alpha^*{}_2}{L_1},$$

$$\bar{R}_{22} = \frac{\alpha^2}{D}\frac{ch\alpha^*{}_1\cos\alpha^*{}_2}{L_1},$$

$$\bar{\bar{R}}_{11} = \bar{\bar{R}}_{21} = \frac{(1-\mu)\,\pi^2\cdot n^2}{2\,Da^2} + \alpha^2\frac{\alpha_2 sh\alpha^*{}_1\cos\alpha^*{}_2 + \alpha^*{}_1 ch\alpha^*{}_1\sin\alpha^*{}_2}{L_2},$$

$$\bar{\bar{R}}_{22} = -\frac{\alpha^2}{D}\frac{sh\alpha^*{}_1\sin\alpha^*{}_2}{L_2},$$

where

$$L_1 = \alpha^*{}_1 sh\alpha^*{}_1\cos\alpha^*{}_2 + \alpha^*{}_2 ch\alpha^*{}_1\sin\alpha^*{}_2,$$

$$L_2 = -\alpha^*{}_1 sh\alpha^*{}_1\sin\alpha^*{}_2 + \alpha^*{}_2 ch\alpha^*{}_1\cos\alpha^*{}_2$$

$$\alpha^*{}_1 = \alpha_1\frac{l}{2},$$

$$\alpha^*{}_2 = \alpha_2\frac{l}{2}.$$

8.2 Finite Differences Equations

Let us consider an m-connectedness regular system that includes only variables from the first group. As an example of such a system, we can use the package of blades with a shroud shown in Fig. 6.1 (the angle between the blades is presumed to be sufficiently small). The blades are presumed to be no extensible. Then the interaction forces between the elements are the following: the normal force and the bending moment from the first group of variables.

The finite-difference equations (6.1) for the variables of the first group $\mathbf{Y}_1$ take the form

$$-\mathbf{FQ}(k-1)+2\mathbf{GQ}(k)-\mathbf{FQ}(k+1)=0, \tag{8.20}$$

where $\mathbf{G}_{11} = \mathbf{G}$, $\mathbf{\Phi}_{11} = \mathbf{F}$; for simplicity's sake, index 2 of vector $\mathbf{Q}$ has been omitted.

Using the results from Chap. 6, the general solution of Eq. (8.20) will take the form

$$\mathbf{Q}(k) = \mathbf{A\Theta}\cos k + \mathbf{B\Theta}\sin k\ (k = 0, 1, \ldots, n), \tag{8.21}$$

$\mathbf{\Theta} = \arccos\left(\mathbf{F}^{-1}\mathbf{G}\right)$ is a square matrix of the order $m \times m$ and $\mathbf{A}$ and $\mathbf{B}$ are m-dimensional vectors of arbitrary constants. The boundary conditions at the input and output of the system are specified for their determination. In this way, it is possible to obtain a solution in analytical form.

In the case of a single-connectedness system ($m = 1$), matrix $\mathbf{\Theta}$ turns into a scalar value $\boldsymbol{\mu}$ and the dynamic compliances will be equal to

$$\left.\begin{aligned}\tilde{F} = \tilde{\Phi} = F\frac{\sin\mu}{\sin n\mu},\\ \tilde{G} = \tilde{T} = F\cos n\mu,\end{aligned}\right\} \tag{8.22}$$

$$\mu = \arccos\frac{\tilde{G}}{F}. \tag{8.23}$$

The finite-difference equations (6.1) for the variables of the second group $\mathbf{Y}_2$ take the form

$$-\mathbf{P}^*\mathbf{Z}(k-1)+2\mathbf{NZ}(k)-\mathbf{P}^*\mathbf{Z}(k+1)=0, \tag{8.24}$$

where

$$\mathbf{P}^* = \begin{bmatrix}-\mathbf{G}^* & \mathbf{S}^*\\ \mathbf{S}^{*\mathbf{T}} & \mathbf{B}^*\end{bmatrix}, \tag{8.25}$$

$$\mathbf{N} = \begin{bmatrix}-\mathbf{G} & \mathbf{S}\\ \mathbf{S}^T & \mathbf{B}\end{bmatrix}, \tag{8.26}$$

$\mathbf{G}^*, \mathbf{S}^*, \mathbf{B}^*$ are determined in (8.9) and (8.9a).

We form a second vector

$$\mathbf{Z}^*(k) = \begin{bmatrix}\mathbf{q}_2(k) & \mathbf{Q}_1(k)\end{bmatrix}^T$$

that contains all antisymmetric components of the coordinates and forces. This vector is expressed through vector $\mathbf{Z}(k)$ with the help of matrices $\mathbf{P}^*$ and $\mathbf{N}$

$$\mathbf{Z}^*(k) = \mathbf{NZ}(k) - \mathbf{P}^*\mathbf{Z}(k+1),\ (k = 0, 1, \ldots, n-1). \tag{8.27}$$

Both matrices $\mathbf{P}^*$ and $\mathbf{N}$ are symmetric, and the matrix in the finite-difference Eq. (8.24) is also "symmetrized" as in Eq. (8.20).

If we set

$$\mathbf{P}^* = \mathbf{F},\ \mathbf{N}{=}\mathbf{D},\ \mathbf{Z}(k) = \mathbf{Q}(k),$$

then all results obtained above for Eq. (8.24) remain valid.

Let us discuss this solution in more detail. We will take the particular solution in the form (6.2.)

$$\mathbf{Q}_2\,[k] = \mathbf{A}_1 C\mathbf{Z}^K, \tag{8.28}$$

$$\mathbf{q}_1\,[k] = \mathbf{A}_2 C\mathbf{Z}^K, \tag{8.29}$$

where $\mathbf{A}_1$ is an l -dimensional vector, $\mathbf{A}_2$ an $m - l$-dimensional vector, and $\mathbf{C}$ an arbitrary constant.

From (8.24)–(8.26) and (8.28)–(8.29) we obtain

$$\left\{\begin{bmatrix} -\mathbf{G} & \mathbf{S} \\ \mathbf{S}^{\mathrm{T}} & \mathbf{B} \end{bmatrix} - \lambda \begin{bmatrix} -\mathbf{G}^* & \mathbf{S}^* \\ \mathbf{S}^{*\mathrm{T}} & \mathbf{B}^* \end{bmatrix}\right\} \begin{matrix} \mathbf{A}_1 \\ \mathbf{A}_2 \end{matrix} = 0, \tag{8.30}$$

where

$$\lambda = \frac{Z + Z^{-1}}{2}, \tag{8.31}$$

and

$$Z_{1,2} = \lambda \pm \sqrt{\lambda^2 - 1}. \tag{8.32}$$

The vector $\mathbf{A} = \begin{bmatrix} \mathbf{A}_1 & \mathbf{A}_2 \end{bmatrix}^T$ is not equal to zero only if

$$\det\left\{\begin{bmatrix} -\mathbf{G} & \mathbf{S} \\ \mathbf{S}^{\mathrm{T}} & \mathbf{B} \end{bmatrix} - \lambda \begin{bmatrix} -\mathbf{G}^* & \mathbf{S}^* \\ \mathbf{S}^{*\mathrm{T}} & -\mathbf{B}^* \end{bmatrix}\right\} = 0.$$

Taking into consideration (8.25) and (8.26), this condition can be written in the form

$$\det\left[\mathbf{N} - \lambda\mathbf{P}^*\right] = 0. \tag{8.33}$$

In this way, we have reached the general eigenvalues problem.

Matrices $\mathbf{N}$ and $\mathbf{P}^*$ are symmetric; therefore, all roots of this equation will be real if matrices $\mathbf{N}$ and $\mathbf{P}^*$ are positively determined. In the opposite case [64], some of the roots of the equation can be complex-conjugated.

Let us examine in more detail how to determine the normal vectors in these cases. Let us assume at the beginning that all roots λ_r ($r = 1,2,\ldots$, m) are different. Then the normal vectors have the following form:

at $|\lambda_r| < 1$ (8.34.a)

$$C_{1r}Z_{1r}^k + C_{2r}Z_{2r}^k = C_{1r}^* \cos k\mu_r + C_{2r}^* \cos k\mu_r, \cos\mu_r = \lambda_r;$$

at $\lambda_r > 1$ (8.34.b)

$$C_{1r}Z_{12}^k + C_{2r}Z_{2r}^k = C_{1r}^* chk\mu_r + C_{2r}^* shk\mu_r, ch\mu_r = \lambda_r;$$

at $\lambda_r < -1$

$$C_{1r}Z_{12}^k + C_{2r}Z_{2r}^k = C_{1r}^* (-1)^k chk\mu_r + C_{2r}^* (-1)^k shk\mu_r, ch\mu_r = -\lambda_r; \tag{8.34.c}$$

at $\lambda_r = 1$

$$C_{1r}Z_{1r}^k + C_{2r}Z_{2r}^k = C_{1r}^* + C_{2r}^* K; \tag{8.34.d}$$

at $\lambda_r = -1$

$$C_{1r}Z_{1r}^k + C_{2r}Z_{2r}^k = C_{1r}^* (-1)^k + C_{2r}^* (-1)^k K. \tag{8.34.e}$$

If among the roots of Eqs. (8.34) there are multiple ones, then the general solution is a linear combination of the corresponding vectors.

Let us consider now the general case where among the roots of Eq. (8.33) there are complex-conjugated ones

$$\lambda_r = a_r + ib_r. \tag{8.35}$$

The normal vectors determined for the given λ_r will also be complex

$$Z_{1r} = \rho_{1r} e^{i\varphi_{1r}}$$
$$Z_{2r} = \rho_{2r} e^{i\varphi_{2r}}.$$

The general solution will contain a linear combination of functions with real coefficients, which has the following form:

$$\left(C_{1r}\rho_{12}^k \cos k\varphi_{1r} + C_{2r}\rho_{2r}^k \cos k\varphi_{2r}\right)\mathbf{A}_r^0 - \left(C_{1r}\rho_{12}^k \sin k\varphi_{1r} + C_{2r}\rho_{2r}^k sikn\varphi_{2r}\right)\mathbf{A}_r^*$$

8.3 Special Types of Boundary Conditions

8.3.1 Nonclosed Systems

Let us go back to Eq. (8.24).

$$-\mathbf{P}^*\mathbf{Z}(k-1) + 2\mathbf{N}\mathbf{Z}(k) - \mathbf{P}^*\mathbf{Z}(k+1) = 0.$$

In the nonclosed systems for its end elements the equality below be fulfilled:

$$\mathbf{Z}(0) = \mathbf{Z}(n) = 0. \tag{8.36}$$

This condition means that all displacements and forces are equal to zero both at the input and the output of the system:

$$\mathbf{Q}_2(0) = \mathbf{Q}_2(n) = \mathbf{q}_1(0) = \mathbf{q}_1(n) = 0 \tag{8.37}$$

From now on we will call these conditions boundary conditions of the first type. We will call the conditions

$$\mathbf{Q}_1(0) = \mathbf{Q}_1(n) = \mathbf{q}_2(0) = \mathbf{q}_2(n) = 0 \tag{8.38}$$

boundary conditions of the second type. The boundary conditions of the first and second types for a system with reflection symmetry elements make it possible to obtain relatively simple solutions. All components of the normal vectors are harmonic functions of the element number.

It can be proven that conditions (8.36) are possible only for those values of ω for which at least one root $|\lambda_r| \leq 1$.

Let some real root $|\lambda_r| = |\lambda| < 1$; we will write the particular solution (8.24) for the given λ in the form

$$\mathbf{Z}[k] = [C_1 \cos k\mu + C_2 \sin k\mu]\,\mathbf{A}, \quad \lambda = \cos\mu.$$

Satisfying the boundary conditions (8.36), we obtain

$$C_1 = 0, \mu = \frac{s\pi}{n}, (s = 0, 1, \ldots, n), \mathbf{Z}(k) = C_2\mathbf{A}\sin\frac{ks\pi}{n}. \tag{8.39}$$

Substituting (8.39) into (8.34) we find that the natural frequencies ω of the system are determined from the equation

$$\det\left[\mathbf{N} - \mathbf{P}^* \cos\frac{s\pi}{n}\right] = 0 \; (s = 1, 2, .., n-1). \tag{8.40}$$

By obtaining this equation, we actually have to solve a new problem, namely: Obtain not the characteristic values, but a series of natural frequencies of the system for every fixed number $s = 1, 2, .., n-1$.

Therefore, the normal vector and the characteristic equation (8.39) and (8.40) can be written in new notation:

$$\left(\mathbf{N} - \mathbf{P}^* \cos\frac{s\pi}{n}\right)\mathbf{C}_s\mathbf{A}^{(s)} = 0, \tag{8.41}$$

$$\mathbf{Z}_s[k] = C_s\mathbf{A}^{(s)} \sin\frac{ks\pi}{n}, \quad (k = 0, 1, 2, \ldots, n;\; s = 1, 2, \ldots, n-1). \tag{8.42}$$

In this way, the characteristic equation of the system at the boundary conditions of the first or the second type has the form

$$\lambda_S = \cos\frac{s\pi}{n}, \; \lambda_S = \lambda_r (r = 1, 2, \ldots, m;\;\; s = 1, 2, \ldots, n-1). \tag{8.43}$$

It is obvious that all characteristic values are in ranges where $|\lambda_r| \leq 1$, even though for at least one value of r.

The normal vectors of the first group are determined in (8.39). The vector of the second group is determined from (8.27)

$$\mathbf{Z}_s^*(k) = \mathbf{N}\mathbf{Z}_s(k) - \mathbf{P}^*\mathbf{Z}_s(k+1).$$

From (8.41) and (8.42) it is possible to obtain

$$\mathbf{Z}_s^*(k) = \mathbf{C}_s^*\mathbf{A}^{*(S)} \cos\frac{ks\pi}{n}, \tag{8.44}$$

where

$$\left.\begin{aligned} C_s^* &= C_s \sin\frac{s\pi}{n} \\ \mathbf{A}^{*(s)} &= \mathbf{P}^*\mathbf{A}^{(s)} \end{aligned}\right\} \tag{8.45}$$

In this way, if vector $\mathbf{Z}_s(k)$ changes following a sine pattern, then vector $\mathbf{Z}_s^*(k)$ changes along a cosine pattern.

Let us point out some properties of the vibration modes of the systems under consideration. Let us rewrite Equality (8.42) in two forms

$$\begin{aligned} \mathbf{Z}_s(k) &= C_s\mathbf{A}^{(s)} \sin\frac{ks\pi}{n}, \\ \mathbf{Z}_s(n-k) &= -C_s\mathbf{A}^{(s)} \cos s\pi \sin\frac{ks\pi}{n}. \end{aligned}$$

It is obvious that at odd values of s

$$\mathbf{Z}_s(n-k) = \mathbf{Z}_s(k), \tag{8.46}$$

and at even values of s

$$\mathbf{Z}_s(n-k) = -\mathbf{Z}_s(k). \tag{8.47}$$

Analogously, at odd s, we obtain from (8.44)

$$\mathbf{Z}_s^*(n-k) = -\mathbf{Z}_s^*(k), \tag{8.48}$$

and at even s

$$\mathbf{Z}_s^*(n-k) = \mathbf{Z}_s^*(k). \tag{8.49}$$

It follows from (8.46)–(8.49) that the vibration modes of a system with respect to its symmetry plane can be divided into two groups:

– Symmetric modes corresponding to an odd value of s when all components of the displacements and internal forces of the first (second) group have the same (different) signs [see expressions (8.46) and (8.48)];
– Skew-symmetric modes corresponding to an even value of s when all components of the displacements and internal forces of the first (second) group have different (the same) signs [see expressions (8.47) and (8.49)].

Let us note also that

$$\lambda(n-s) = -\lambda(s). \tag{8.50}$$

8.3.2 Closed Systems

Another kind of very simple system with reflection symmetry elements is a closed system for which the periodicity condition is fulfilled.

Let us write again Eq. (8.24) for a system of n elements.

$$-\mathbf{P}^*\mathbf{Z}(k-1) + 2\mathbf{N}\mathbf{Z}(k) - \mathbf{P}^*\mathbf{Z}(k+1) = 0.$$

The periodicity condition means that

$$\mathbf{Z}(n) = \mathbf{Z}(0). \tag{8.51}$$

The particular solutions of the two types satisfy Eq. (8.24) and condition (8.51)

$$\mathbf{Z}_s(k) = C_s\mathbf{A}^{(s)} \sin\frac{2ks\pi}{n} \tag{8.52}$$

or

$$\mathbf{Z}_{1s}(k) = C_{1s}\mathbf{A}^{(s)} cos\frac{2ks\pi}{n} \tag{8.53}$$

Correspondingly,

$$\mathbf{Z}_s^*(k) = C_{1s}\mathbf{A}_1^{(s)} \cos\frac{2ks\pi}{n} = \mathbf{Z}_{1s}(k), \tag{8.52a}$$

$$\mathbf{Z}_{1s}^*(k) = C_s\mathbf{A}^{(s)} \sin\frac{2ks\pi}{n} = \mathbf{Z}_s(k). \tag{8.53a}$$

After substituting (8.52) or (8.53) into (8.24), we obtain the characteristic equation

$$\det\left[\mathbf{N} - \mathbf{P}^* \cos\frac{2s\pi}{n}\right] = 0. \tag{8.54}$$

The characteristic values are determined from the condition

$$\lambda_r = \lambda_s, \text{ where } `\div\text{'e } \lambda_s = \cos\frac{2s\pi}{n} \quad (s = 0, 1, \ldots, n-1), \tag{8.55}$$

and the values $|\lambda_r| < 1 \quad (r = 1, 2 \ldots, m)$ are roots of the characteristic equation

$$\det\left[\mathbf{N} - \lambda_r \mathbf{P}^*\right] = 0. \tag{8.56}$$

Since $\lambda_{(n-s)} = \lambda_s$, the corresponding characteristic values (and therefore the natural frequencies) are multiple [20, 45, 46]. However, the normal modes corresponding to them are different and linearly independent. Modes (8.52) and (8.52a) correspond to one of the multiple characteristic values, while modes (8.53) and (8.53a) correspond to the other one.

At s=0 и s=n /2 (n – even)

$$\lambda_0 = 1 \quad \text{and} \quad \lambda_{n/2} = -1.$$

The normal modes are determined by the equations:

$$\begin{aligned}
&\text{at } s = 0\\
&\mathbf{Z}(k) = C_{10}A^{(0)}; \qquad \mathbf{Z}^*(k) = 0,\\
&\text{at } s = n/2\\
&\mathbf{Z}(k) = (-1)^k C_{1n/2}\mathbf{A}^{(n/2)}; \qquad \mathbf{Z}^*(k) = 0.
\end{aligned}$$

Let us point out some general properties of open systems with boundary conditions of the first and second types for nonclosed and closed systems:

1. The natural frequencies of skew-symmetric normal modes for a nonclosed system with boundary conditions of the first type coincide with the natural frequencies of a closed system. The frequencies of the symmetric normal modes of such a system coincide with the natural frequencies of a closed system.
2. Two linearly independent normal modes correspond to the multiple frequencies of a closed system. The first of these modes coincides with an analogous skew-symmetric mode of a nonclosed system under boundary conditions of the first type. The second of these modes coincides with an analogous skew-symmetric mode for a nonclosed system under boundary conditions of the second type.
3. The natural frequencies of a nonclosed system under the boundary conditions of the first type with symmetric normal modes are absent in a closed system. Analogously, the natural frequencies of a nonclosed system under the boundary conditions of the second type with skew-symmetric normal modes are absent in a closed system.
4. The nonmultiple natural frequencies of a closed system, which correspond to $s = 0$ and $s = n/2$ (n is even), coincide with the trivial series of natural frequencies of a nonclosed system.

5. In a closed system composed of $2n$ elements, the frequency spectrum includes all frequencies of a nonclosed system composed of n elements.

Let us consider an example at n=6

a. Nonclosed system $\lambda_s = \cos \dfrac{\pi s}{6}$

$$\lambda_0 = 1,\ \lambda_{1,5} = \pm\frac{\sqrt{3}}{2},\ \lambda_{2,4} = \pm\frac{1}{2},\ \lambda_3 = 0, \lambda_6 = -1;$$

b. Closed system $\lambda_s^* = \cos \dfrac{2\pi s}{6}$

$$\lambda^*{}_0 = 1,\ \lambda^*{}_1 = \lambda_s^* = \frac{1}{2},\ \lambda^*{}_2 = \lambda^*{}_4 = -\frac{1}{2},\ \lambda^*{}_3 = -1;$$
$$\lambda^*{}_0 = \lambda_0,\ \lambda^*{}_1 = \lambda_s^* = \lambda_1,\ \lambda^*{}_2 = \lambda^*{}_4 = \lambda_2,\ \lambda_3 = \lambda_6.$$

8.4 Filtering Properties of System with Reflection Symmetry Elements

Let us begin with the consideration of a single-connectedness system ($m = 1$) and write Eq. (8.24) for this system:

$$-P^*Z(s-1) + 2NZ(s) - P^*Z(s+1) = 0.$$

where P^* and N are now scalar values, not matrices, s is element number.

Either the displacements or the force can be selected as $Z(s)$. In the first case, equations of the dynamic stiffness method and in the second case equations of the dynamic compliances method are obtained. Here we can speak about a travelling wave for an infinite system if for $Z(s)$ there is a solution in the form

$$Z(s,t) = A \exp\left[i(\omega t - s\mu)\right], \tag{8.57}$$

where

$$\mu = \pm \arccos \lambda, \tag{8.58}$$

$$\lambda = \frac{N}{P^*} \tag{8.59}$$

It can be seen from (8.58) that the following condition has to be fulfilled:

$$|\lambda| < 1,\ \left|\frac{N}{P^*}\right| < 1. \tag{8.60}$$

This condition ensures the passage of the waves through the system. Therefore, one can obtain from this inequality the frequency ranges for which the formation of travelling waves, i.e., pass bands, is possible.

Let us now examine an m-connectedness system. It has been shown in [80] that the pass bands for an m-connectedness system coincide with a domain where a solution of the type (8.57) is possible for one or several pairs of component vectors $\mathbf{Z}(s)$. Therefore, in order for waves to pass at a given frequency ω, it is necessary that at least one of the characteristic values λ_r $(r = 1, 2, \ldots, m)$ of Eq. (8.33), calculated at a frequency ω, satisfy the condition

$$|\lambda_r| < 1. \tag{8.61}$$

8.5 Numerical Examples

8.5.1 Single-Connectedness Systems

Example 1. *A chain system consists of concentrated masses equal to m connected with each other by springs with stiffness C (Fig. 8.5a). The value of the first and last masses is equal to* m. *Let us designate the number of the masses by n + 1. An element of the system (Fig. 8.5б) consists of two masses* m *connected by a spring. The number of elements is equal to* n. *In order to determine the unknown forces that act between the elements, we use the dynamic compliances method (8.20).*

The dynamic compliance of an element is

$$N = -\frac{1}{2m\omega^2}\frac{1-\dfrac{\omega^2}{p^2}}{1-\dfrac{\omega^2}{p^*}}, P^* = -\frac{1}{2m\omega^2}\frac{1}{1-\dfrac{\omega^2}{p^*}},$$

where

$$p^{*2} = \frac{C}{m}, p^2 = \frac{2C}{m}.$$

The characteristic value is

$$\lambda = 1 - \alpha, \alpha = \frac{m\omega^2}{C},$$

$$|\lambda| < 1 \text{ at} 0 < \alpha < 2 \text{ (Fig.8.5c)}.$$

The natural frequency is

$$\omega_s^2 = \frac{C}{m}\left(1 - \cos\frac{s\pi}{n}\right) (s = 1, 2, \ldots, n-1). \tag{8.62}$$

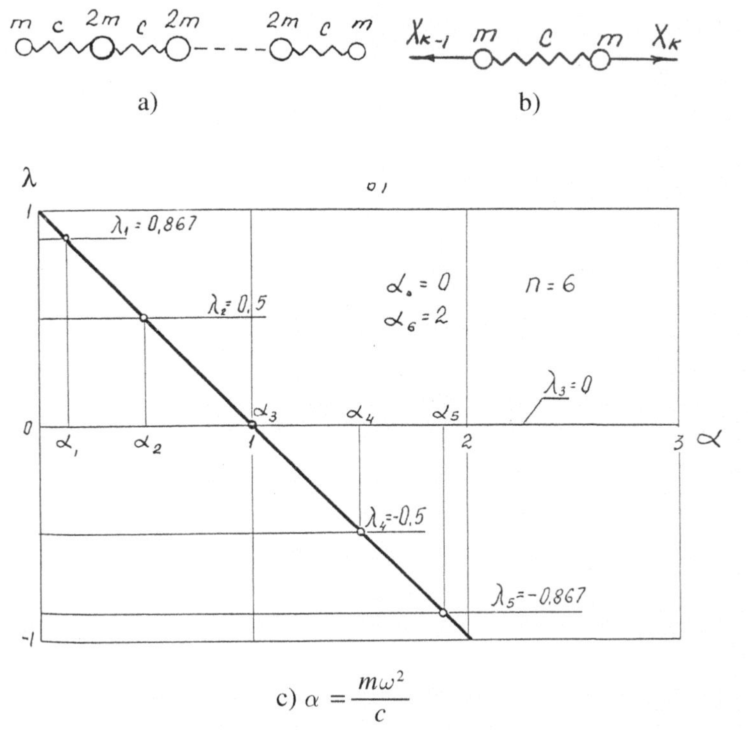

Fig. 8.5 A single-connectedness chain system with reflection symmetry elements (**a**), a regular element (**b**), and characteristic values (**c**)

Two more trivial frequencies at which $Z(s) = 0$ and the frequencies of the system coincide with the frequency of the element are possible:

$$\omega_0 = 0, \alpha_0 = 0, \; Z^*(s) = U(s) = a,$$

$$\omega_n^2 = 2\frac{C}{m}, \; \alpha_n = 2, \; Z^*(s) = U_s = (-1)^s a;$$

where U_s is the displacement of the mass with number s and a is a some constant.

The entire frequency spectrum is in the range $0 \leq \alpha \leq 2$.

The entire system of $n+1$ masses (n elements) has $n+1$ frequencies, including the zero frequency.

Let us find all frequencies at $n = 6$.

The trivial frequencies are $\omega = 0, \omega_6^2 = 2\frac{C}{m}$.

From (8.62)

$$\omega_1^2 = 0.143\frac{C}{m}; \omega_2^2 = 0.5\frac{C}{m}; \omega_3^2 = \frac{C}{m}; \omega_4^2 = 1.5\frac{C}{m}; \omega_5^2 = 0.857\frac{C}{m}.$$

For an infinite system $n \to \infty$, the pass band also corresponds to

$$|\lambda| < 1 \quad 0 < \omega < \sqrt{\frac{2C}{m}}.$$

The phase velocity of the wave is $\upsilon = \frac{l\omega}{\mu}$, where l is the distance between the masses, $\mu = \arccos \lambda$;

$$\text{at } \omega \to \sqrt{\frac{c}{2m}} \quad \upsilon \to \upsilon_{\min} = \frac{2\,l}{\pi}\sqrt{\frac{c}{2m}}$$

Let us now close the system, i.e., implement the requirement $Z_n = Z_0$. The elements of the system do not change, so that the expression for λ also will not change; however, the frequency equation will be different:

$$\tilde{\omega}_s^2 = \frac{C}{m}\left(1 - \cos\frac{2s\pi}{n}\right)$$
$$s = 0, 1, \ldots, \frac{n}{2} \text{ at even } n$$
$$s = 0, 1, \ldots, \frac{n-1}{2} \text{ at odd } n.$$

All frequencies, except for ω_0 and $\omega_{n/2}$, are multiple, and they are included in the spectrum of the nonclosed system (8.62), $\tilde{\omega}_0 = 0 = \omega_0$. At even n, $\tilde{\omega}_{n/2} = 2\frac{C}{m} = \omega_n^2$; at odd n, this value of the frequency is absent.

Example 2. *A multispan regular weightless beam with a mass in the middle of each span (Fig. 6.4a), the number of elements* n *is equal to the number of masses.*

$$\lambda = 2\frac{1 - 7\alpha}{1 + 2\alpha}, \quad \alpha = \frac{ml^3\omega^2}{768EJ} \tag{8.63}$$

l is the length of the span, EJ the bending stiffness of the beam. The natural frequencies are

$$\omega_{\min}^2 = \omega_0^2 = \frac{48EJ}{ml^3} \text{ at } \alpha_0 = \frac{1}{16}.$$

The entire frequency spectrum is in the range

$$\frac{1}{16} < \alpha \leq \frac{1}{4}.$$

The limit unreachable value of the frequency at $\alpha_n = 0.25$ is equal to $\omega_n = \frac{192EJ}{ml^3}$. This value corresponds to the vibration of a beam with fixed ends.

For a closed system

$$\tilde{\omega}_s^2 = \frac{384EJ}{ml^3}\frac{2-\cos\dfrac{2s\pi}{n}}{7+\cos\dfrac{2sS\pi}{n}};$$

$$s = 0, 1, 2, \ldots, \frac{n-1}{2} \quad \text{(for odd } n\text{)},$$

$$s = 0, 1, 2, \ldots, \frac{n}{2} \quad \text{(for even } n\text{)}.$$

$$\text{At } s = 0,\ \tilde{\omega}_0^2 = \frac{192EJ}{ml^3}, \alpha_0 = \frac{1}{16};$$

$$\text{At } s = \frac{n}{2},\ \tilde{\omega}_{\frac{n}{2}}^2 = \frac{48EJ}{l^3} = \omega_0^2, \alpha_{\frac{n}{2}} = \frac{1}{4}.$$

Example 3. *A regular multispan beam with distributed mass (Fig. 8.6a). Element of the system – a single-span beam (Fig. 8.6b), the number of elements is equal to the number of supports minus one.*

The dynamic compliances of the element are

$$\left.\begin{aligned} d = t &= \frac{l}{EJ\alpha}\frac{\sin\alpha ch\alpha - \cos\alpha sh\alpha}{2\sin\alpha ch\alpha} \\ f &= \frac{l}{EJ\alpha}\frac{sh\alpha - \sin\alpha}{2\sin\alpha sh\alpha} \\ \alpha^4 &= \frac{m_0\omega^2 l^4}{EJ}, \end{aligned}\right\} \tag{8.64}$$

where EJ is the bending stiffness in the plane of the vibrations, l is the length of the span, and m_0 is the mass of a unitary length of the beam.

According to (5.15)

$$\lambda = \frac{d}{f}.$$

After substituting expression (8.64), we obtain

$$\lambda = \frac{ch\alpha\sin\alpha - \cos\alpha sh\alpha}{sh\alpha - \sin\alpha}. \tag{8.65}$$

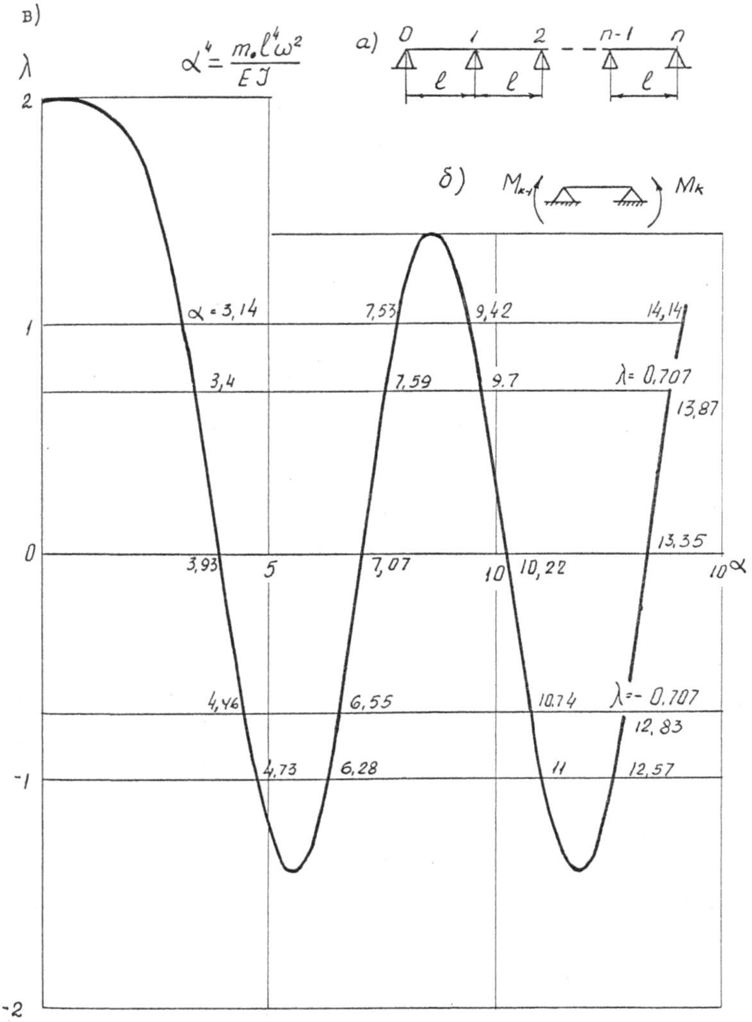

Fig. 8.6 A regular multispan beam (**a**), a regular element (**b**), and characteristic values (**c**)

The values $\lambda = \lambda(\alpha)$ are charted in Fig. 8.6c. The natural frequencies correspond to the values of α determined from the equation

$$\cos\frac{\pi S}{n} = \lambda\,(s = 1, 2, \ldots, n-1)\,.$$

The trivial series of the frequencies corresponding to the vibration frequency of a single element can be obtained from the equations [106]

$$sh\alpha_r \sin\alpha_r = 0$$

or

$$\alpha_r = \pi r\,(r = 1,2,\ldots)\,.$$

It can be seen from Fig. 8.6c that the natural frequencies, without to the dependence on the number of elements n, are located in certain intervals where $|\lambda| \leq 1$.

The first interval, or the so-called "first zone of frequency concentration," is located in the range $3.14 \leq \alpha \leq 4.73$, the second zone in the range $6.28 \leq \alpha \leq 7.53$, the third zone in the range $9.42 \leq \alpha \leq 1$, the fourth zone in the range $12.57 \leq \alpha \leq 14.14$, etc. The travelling waves in these zones is possible at $n \to \infty$. The frequencies corresponding to α= 3.14, 6.28, 9.42, 12.57 ... are frequencies of a trivial series coinciding with the frequencies of a single beam.

The vibration frequencies corresponding to α =4.73, 7.53, 11, 14.14 ... are the frequencies of a beam with its two ends fixed. For each zone, these frequencies are extreme and unreachable.

Let us agree to provide α with two indices α_{s2}. The first index is the number of the frequency in the given zone, while the second one is the number of the zone. Let the number of spans in the system be $n = 4$.

$$\lambda = \cos\frac{\pi s}{4}\,(s = 1,2,3)\,.$$

The trivial series is

$$\alpha_{01} = 3.14;\ \alpha_{02} = 6.28;\ \alpha_{03} = 9.42;\ \alpha_{04} = 12.57;\ \ldots$$
$$s = 1,\ 3,\ \lambda = \pm 0.707;$$
$$\alpha_{11} = 3.4;\ \alpha_{12} = 6.55;\ \alpha_{13} = 9.7;\ \alpha_{14} = 12.83;\ldots$$
$$\alpha_{31} = 4.46;\ \alpha_{32} = 7.59;\ \alpha_{33} = 10.22;\ \alpha_{34} = 13.35;\ldots$$
$$s = 2,\ \lambda = 0;$$
$$\alpha_{21} = 3,93;\ \alpha_{22} = 7,07;\ \alpha_{23} = 10,73;\ \alpha_{34} = 13,87;\ldots$$

All obtained frequencies will be included as well in the spectrum of a system where the number of elements is divisible by four.

For a closed system, the spectrum $\alpha = 4.73,\ 7.53,\ 11,\ 14.14,\ldots$ will already be reachable. Let us note that the condition $\lambda = \pm 1$ is equivalent to the two equations

$$sh\alpha \sin\alpha = 0$$

and

$$ch\alpha \cos\alpha = 1,$$

from which the frequencies of a freely supported and a fixed beam can be obtained.

8.5.2 Two-Connectedness Systems

Example 4. *A single-span weightless beam with concentrated masses (Fig. 8.7). The number of masses is n − 1; on the supports we place a mass of* $\frac{m}{2}$. *An element of the system is shown in Fig. 8.7b. The number of elements is* n. *We select* $\mathbf{Z}(s)$ *as the vector*

$$\mathbf{Z}(s) = \begin{bmatrix} M(s) \\ V(s) \end{bmatrix} (k = 0, 1, 2, \ldots, n).$$

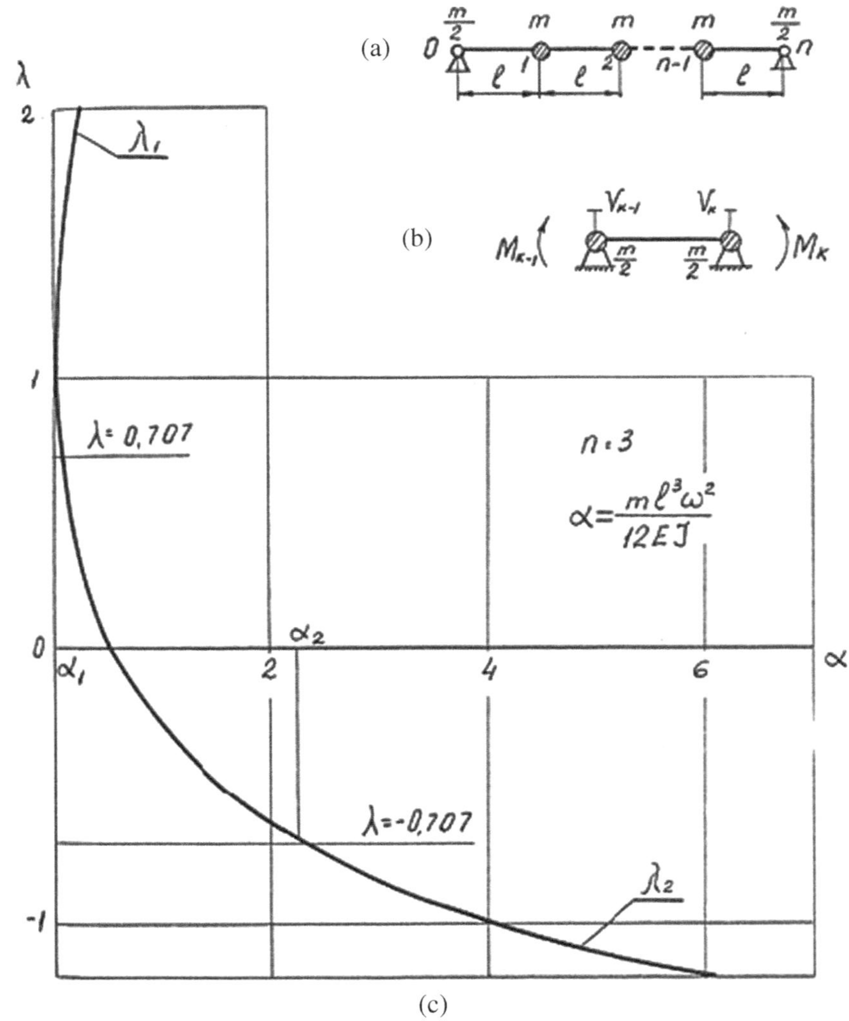

Fig. 8.7 A single-span weightless beam with concentrated masses (**a**), a regular element (**b**), and characteristic values of the mixed matrix (**c**)

where $M(s)$ is the bending moment, $V(s)$ the vertical displacement of the mass with the number s. The elements of the mixed matrices $\mathbf{P}^*$ and $\mathbf{N}$ (8.25) and (8.26) are:

$$G = \frac{l}{3EJ}, S = \frac{1}{l}, B = -\frac{m}{2}\omega^2,$$
$$G^* = -\frac{l}{6EJ}, S^* = -\frac{1}{l}, B^* = 0.$$

Equation (8.33) takes the form

$$\begin{vmatrix} \frac{l}{3EJ} + \frac{l}{6EJ}\lambda & -\frac{1}{l} + \frac{1}{l}\lambda \\ -\frac{1}{l} + \frac{1}{l}\lambda & \frac{m\omega^2}{2} \end{vmatrix} = 0.$$

After expanding the determinant, we find the solution from the quadratic equation

$$\lambda_{1,2} = 1 + \frac{\alpha}{2} \pm \sqrt{3\alpha + \frac{\alpha^2}{4}}, \ \alpha = \frac{ml^3\omega^2}{12EJ}.$$

The curves representing the two characteristic values $\lambda_{1,2}$ as a function of α are charted in Fig. 8.1c. As can be seen from Fig. 8.1c: $\lambda_1 > 1, |\lambda_2| < 1$ at $0 < \alpha < 4$. These values determine the pass band for an endless system.

By setting $\lambda_2 = \cos\frac{\pi s}{n}$, we obtain a relation for determining the frequency:

$$\omega_s^2 = \frac{12EJ}{ml^3}\frac{\left(1 - \cos\dfrac{s\pi}{n}\right)}{2 + \cos\dfrac{s\pi}{n}}.$$

Example 5. Torsion-longitudinal vibrations of a crankshaft. *Let us consider the torsion-longitudinal vibrations of a crankshaft [56]. The model of the shaft is shown in Fig. 8.8. This is a crankshaft with* $(n-1)$ *disks, where* n *is the number of knee pieces. These disks are details from the rod-piston group. Each disk has a mass* m_s *and polar moment of inertia* Q_s $(s = 0, 1, \ldots, n)$. *The disks are connected to each other by elastic inertia-free connections each one having longitudinal* (C_{11}), *torsion* (C_{22}), *and torsion-longitudinal* (C_{12}) *stiffness. The regular element of the shaft* $m_s = m$, $Q_k = Q$ *is presented in Fig. 8.8b. The number of elements is* n.

We select $\mathbf{Z}(s)$ as the vector

$$\mathbf{Z}(s) = \begin{bmatrix} M(s) \\ V(s) \end{bmatrix} (s = 0, 1, 2, \ldots, n),$$

where $M(s)$ is the bending moment and $V(s)$ the horizontal displacement of the mass with number s.

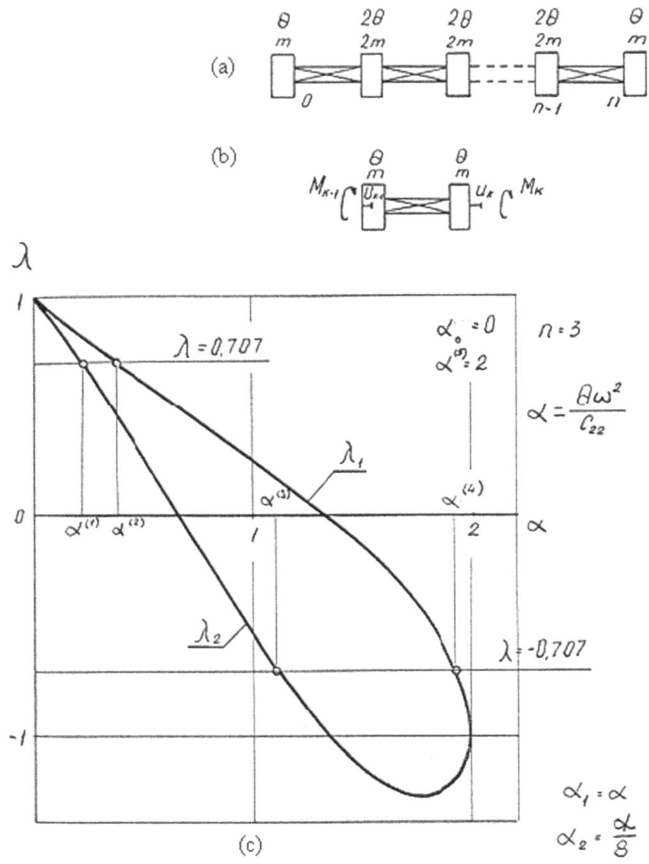

Fig. 8.8a–c Torsion-longitudinal vibrations of a crankshaft (**a**), a regular element (**b**), and characteristic values of a mixed matrix (**c**)

In this example, the elements of mixed matrices $\mathbf{P}^*$ and $\mathbf{N}$ (8.25) and (8.26) have the following form:

$$G = -\frac{1}{2\theta\omega^2}\frac{1-\dfrac{\theta\omega^2}{C_{22}}}{1-\dfrac{\theta\omega^2}{2C_{22}}},\ S = S^* = -\frac{C_{11}}{C_{12}},$$

$$B = C_{11}\left(1-\frac{m\omega^2}{C_{11}}\right),\ G^* = \frac{1}{2\theta\omega^2}\frac{1}{1-\dfrac{\theta\omega^2}{2C_{22}}},\ B^* = C_{11}.$$

Equations (8.33) get the form:

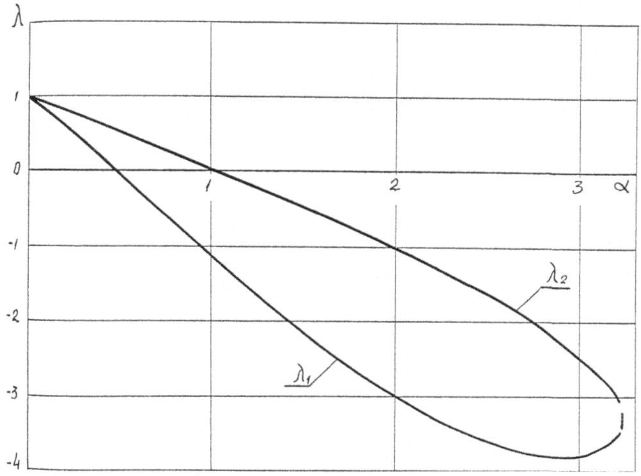

Fig. 8.9 Characteristic values of the mixed matrix of a system element (Fig. 8.8b). ($\alpha_1 = 2\alpha,\ \alpha_2 = \alpha/32$)

$$\det\begin{vmatrix} G - \lambda G^* & -S + \lambda S^* \\ -S + \lambda S^* & -B + \lambda B^* \end{vmatrix} = 0$$

Its roots are

$$\lambda_{1,2} = 1 - \frac{\alpha + \alpha_1 \pm \sqrt{(\alpha - \alpha_1)^2 + 4\alpha^2 (2 - \alpha)}}{2\,(1 - 2\alpha_2 + \alpha\alpha_2)}$$
$$\alpha = \frac{\theta\omega^2}{C_{22}},\ \alpha_1 = \frac{m\omega^2}{C_{11}},\ \alpha_2 = \frac{\theta\omega^2 C_{11}}{C_{12}^2}. \tag{8.65}$$

The natural frequencies are determined from the equation

$$\lambda = \cos\frac{s\pi}{n}\ (s = 1, 2, \ldots, n - 1).$$

There are two trivial frequencies at $\alpha_0 = 0$ and $\alpha_n = 2$.

Figures 8.8c, 8.9, and 8.10 present the functional dependencies $\lambda_1 = \lambda_1(\alpha)$ and $\lambda_2 = \lambda_2(\alpha)$ at various ratios between the coefficients α, α_1, and α_2. The wave pass band is at $n \to \infty 0 \le \alpha \le 2$. At $\alpha > 3.2$ (Fig. 8.9–8.10b), the roots of Eq. (8.65) turn out to be complex-conjugated [see (8.3)].

8.5.3 Three-Connectedness System

Example 6. *A frame consisting of* n *weightless round arcs (Fig. 8.11a). Masses* m *are mounted at the sections where the arcs connect to each other. The frame has two hinged movable supports where masses* $\frac{m}{2}$ *are concentrated. The system*

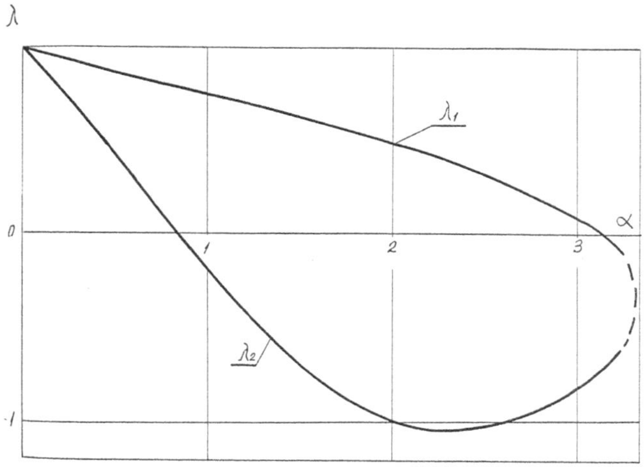

Fig. 8.10 Characteristic values of the mixed matrix of a system element (Fig. 8.8b). $\alpha_1 = \alpha/4$, $\alpha_2 = \alpha/8$)

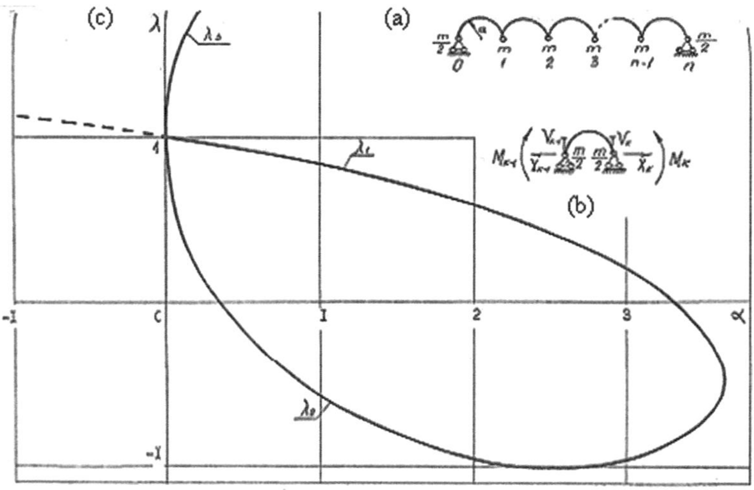

Fig. 8.11 A three-connectedness system **(a)**, a regular element **(b)**, and characteristic values of the mixed matrix **(c)**

satisfies the boundary conditions of the first type. Connections are imposed only on the vertical displacements V_0 *and* V_n *of the end sections. Let us note that the boundary conditions of the second type are satisfied by a structure in which the horizontal displacements and the rotation angles in the end sections are excluded simultaneously. The regular element is shown in Fig. 8.11b. The number of elements is* n.

We select $\mathbf{Z}(s)$ as the vector

$$\mathbf{Z}(s) = [M(s), X(s), V(s)]^T ,$$

where $M(s)$ is the bending moment, $X(s)$ the normal force, and $V(s)$ the vertical displacement ($s = 0, 1, \ldots, n$).

The characteristic equation

$$\det \begin{vmatrix} G_{11} - \lambda G_{11}^* & G_{12} - \lambda G_{12}^* & -S_{11} - \lambda S_{11}^* \\ G_{21} - \lambda G_{21}^* & G_{22} - \lambda G_{22}^* & -S_{12} - \lambda S_{12}^* \\ -S_{11} + \lambda S_{11}^* & -S_{12} + \lambda S_{12}^* & -B + \lambda B^* \end{vmatrix} = 0,$$

where the elements of the mixed matrices $\mathbf{P}^*$ and $\mathbf{N}$ are:

$$G_{11} = \frac{3}{8}\frac{\pi r}{EJ}\frac{1 - 0,181\alpha}{1 - 0,393\alpha}, \quad G_{11}^* = \frac{\pi r}{8EJ}\frac{1 + 0,243\alpha}{1 - 0,393\alpha},$$

$$G_{12} = G_{21} = \frac{r^2}{EJ}\frac{1}{1 - 0,393\alpha}, \quad G_{12}^* = -\frac{r}{EJ}\frac{1}{1 - 0,393\alpha},$$

$$G_{22} = -\frac{1}{m\omega^2}\frac{1 - 0,786\alpha}{1 - 0,393\alpha}, \quad G_{22}^* = \frac{1}{m\omega^2}\frac{1}{1 - 0,393\alpha},$$

$$S_{11} = S_{11}^* = \frac{1}{2r}, \; S_{12} = S_{21} = S_{12}^* = S_{21}^* = 0,$$

$$B = -\frac{m\omega^2}{2}, B^* = 0, \; \alpha = \frac{m\omega^2 r^3}{EJ},$$

r is the arc radius and EJ is the bending stiffness.

Expanding this determinant, we obtain

$$\alpha^2 (0.387\beta - 0.47) - 3.14\alpha\beta + \beta^3 = 0, \; \beta = 1 - \lambda.$$

All three characteristic values as a function of α are given in Fig. 8.11c, from where it can be seen that in the interval $0 < \alpha < 3.62$:

$$\lambda_3 > 0, |\lambda_1| < 0 \text{ and } |\lambda_2| < 0.$$

Two trivial frequencies correspond to $\alpha_0 = 0$ and $\alpha_n = 2.54$. The remaining values are determined from the equation

$$\lambda_{1,2} = \cos\frac{s\pi}{n} \; (s = 1, 2, \ldots, n - 1)$$

in the same way as in examples 4 and 5.

For example, at $n = 2$, the system has four degrees of freedom and, correspondingly, four frequencies. Two frequencies are shown above, $\alpha_0 = 0$ and $\alpha_3 = 2.54$.

Let us find the remaining two frequencies. At $n = 2$, $s = 1$, $\lambda_1 = \lambda_2 = 0$. The curves $\lambda_1 = \lambda_1(\omega)$ and $\lambda_2 = \lambda_2(\omega)$ cross the abscissa coordinate at points $\alpha_1 = 0.35$, $\alpha_4 = 3.35$ (Fig. 8.11c).

8.6 Systems Consisting of Skew-Symmetric (Antisymmetric) Elements

Let us consider the skew-symmetric element shown in Fig. 8.12.

Let us assume that this element executes vibrations with a mode that is skew-symmetric with respect to plane Π. Then part of the coordinates included in vector $\mathbf{q}_a$ will coincide in value and sign with the corresponding coordinates included in vector $\mathbf{q}_b$. The remaining coordinates are also equal to each other but have opposite signs. As we did in the case of a reflection symmetry element, let us specify that all coordinates with the same sign, as well as the internal forces corresponding to them, will be placed in the first group. The remaining generalized coordinates and the internal forces corresponding to them will belong to the second group. It is obvious that the picture will be just the opposite in the case where the vibrations modes of the element are symmetric with respect to plane Π.

Let us consider an m-connectedness system consisting of sequentially conjugated skew-symmetry elements and write the mixed matrix for this system

$$-\mathbf{P}^*\mathbf{Z}(s-1) + 2\mathbf{N}\mathbf{Z}(s) - \mathbf{P}^*\mathbf{Z}(s+1) = 0, \ (s = 1, 2, \ldots, n-1). \qquad (8.24)$$

Let us include in vector $\mathbf{Z}(s)$ the components of the coordinates and forces related only to the first or only to the second group. Then all results obtained for a system consisting of reflection symmetry elements will also remain valid for systems consisting of skew-symmetric elements. In particular, under the boundary conditions $\mathbf{X}(0) = \mathbf{X}(n) = 0$, the characteristic values of the system are determined from the analogous equation:

$$\det\left[N - \lambda P^*\right] = 0, \qquad (8.66)$$

where

$$\lambda = \cos\frac{s\pi}{n} \ (s = 1, 2, \ldots, n-1). \qquad (8.67)$$

Let us consider the following example.

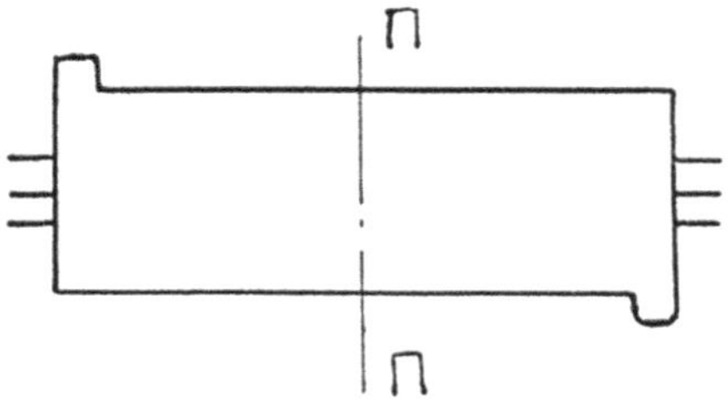

Fig. 8.12 Skew-symmetric element

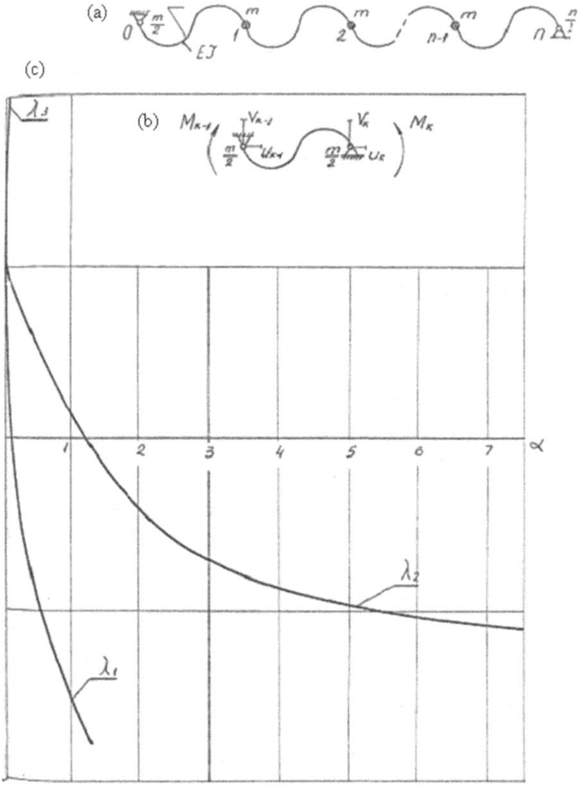

Fig. 8.13 A three-connectedness system (**a**), a regular skew-symmetry element (**b**), and characteristic values of the mixed matrix (**c**)

We have a weightless curvilinear frame with concentrated masses m that is mounted on two hinged-motionless supports (Fig. 8.13a). The element of the frame shown in Fig. 8.13b is skew-symmetric. The first group of coordinates and forces includes the rotation angles of the element end sections and the normal and transversal forces N and Q. The second group includes the horizontal and vertical coordinates u and v, as well as the bending moments M. The boundary conditions of the frame, at which Equality (8.67) is fulfilled, correspond either to hinged-motionless supports or to supports impeding the rotation of the end element. Now we consider the first case. In the same way as in example 8.5, we add at the ends of the frame a mass with a value of $\frac{m}{2}$. We select as $\mathbf{Z}(s)$ the following vector:

$$\mathbf{Z}(s) = [M(s), U(s), V(s)]^T, \ (s = 0, 1, \dots, n).$$

The elements of mixed matrices $\mathbf{P}^*$ and $\mathbf{N}$ are:

$$G_{11} = 1,84\frac{r}{EJ}, G_{11}^* = -1,3\frac{r}{EJ},$$

$$S_{11} = S_{11}^* = \frac{1}{4r}, S_{12} = S_{21} = S_{12}^* = S_{21}^* = -0,318\frac{1}{r},$$

$$B_{11} = -\frac{EJ}{\pi r^3}, \alpha = B^*_{22}, B^*_{22} = B_{12} = B^*_{12} = B_{21} = B^*_{21} = 0,$$
$$B_{22} = \frac{EJ}{\pi r^3}(1-\alpha),$$

where

$$\alpha = \frac{\pi}{2}\frac{m\omega^2 r^3}{EJ},$$

r is the radius of the arc, EJ is the bending stiffness.

The characteristic equation is

$$\det \begin{vmatrix} G_{11} - \lambda G^*_{11} & -S_{11} + \lambda S^*_{11} & -S_{12} + \lambda S^*_{12} \\ S_{11} + \lambda S^*_{11} & -B_{11} + \lambda B^*_{11} & -B_{12} + \lambda B^*_{12} \\ -S_{12} + \lambda S^*_{12} & -B_{12} + \lambda B^*_{12} & -B_{22} + \lambda B^*_{22} \end{vmatrix} = 0.$$

After expansion, we obtain the equation

$$16\alpha^2 + \left(16\alpha + 6,64\alpha^2\right)\beta + 4\alpha\beta^2 - \beta^3 = 0, \ \beta = \lambda - 1.$$

The curves $\lambda_r(\alpha) \quad (r = 1, 2, 3)$ are represented on Figs. 8.13c and 8.14.

There are no trivial frequencies series in a system consisting of skew-symmetry elements.

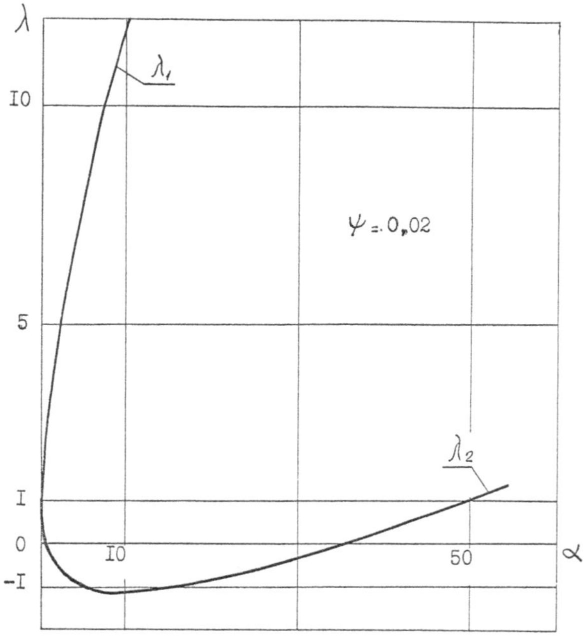

Fig. 8.14 Characteristic values of the mixed matrix

The natural frequencies are determined from the equation

$$\lambda_{2,3} = \cos\frac{s\pi}{n} \quad (s = 1, 2, \ldots, n-1)\,.$$

For example, at $n = 2,\ s = 1\ \cos\dfrac{\pi}{2} = 0.$

$$\alpha_1 = 0.0909,\ \alpha_2 = 1.19.$$

These values of α correspond to frequencies belonging to the spectrum of the system at any even number of elements (odd number of masses).

The characteristic values for a three-connectedness system with hinged supports can also be obtained analogously.

For a closed system consisting of skew-symmetry elements, the phase shift for the waves between the components of the forces and the displacements is equal to zero or is a multiple of $\frac{\pi}{2}$. Earlier [45, 102], this fact was mentioned only for systems with reflection symmetry elements.

Chapter 9
Self-Similar Structures

9.1 Introductory Part: Examples of Self-Similar Mechanical Structures

Self-similar structures are often used in real-life engineering structures. This chapter is devoted to the vibrations of such structures. Figure 9.1a–e shows examples of such mechanical systems.

The simplest example of such a system is the step-up rod, which performs longitudinal vibrations. The lengths of the rod sections are identical and the cross-section areas change in the same proportion from section to section (Fig. 9.1a). The second example is a shaft executing bending vibrations (Fig. 9.1b). The masses and the equatorial moments of the disks as well as the static rigidities of the inertia-free connections change proportionally from section to section. The more complex self-similar systems include systems in which the disks are connected by conical shells (Fig. 9.1c). Figure 9.1d shows a section of a drum-type compressor rotor with a conical shell. The spatial crankshaft (Fig. 9.1e) [53, 56] can also be classified with structures of this type (Chap. 6). Here every consecutive section is formed from the preceding one by a shift of the constant vector followed by a rotation around it at some constant angle. More complex systems can be obtained if a contraction or expansion of the element takes place simultaneously with the shift. Such elements have the property of geometric similarity (homothetic).

The main idea in the calculation of self-similar systems, just as in the case of regular systems, consists in the fact that it is possible to find the dynamic characteristics of assembled systems from the known dynamic characteristics of a single standard element.

Let e_k be the dynamic compliance matrix for the element number k of the system. If the below condition is satisfied for every k, then

$$\mathbf{e}_{k+1} = \gamma \mathbf{e}_k, \tag{9.1}$$

where γ is a scalar quantity. Condition (9.1) is satisfied by systems whose elements are geometrically similar [57, 58].

L.Y. Banakh, M.L. Kempner, *Vibrations of Mechanical Systems with Regular Structure*, Foundations of Engineering Mechanics, DOI 10.1007/978-3-642-03126-7_9,

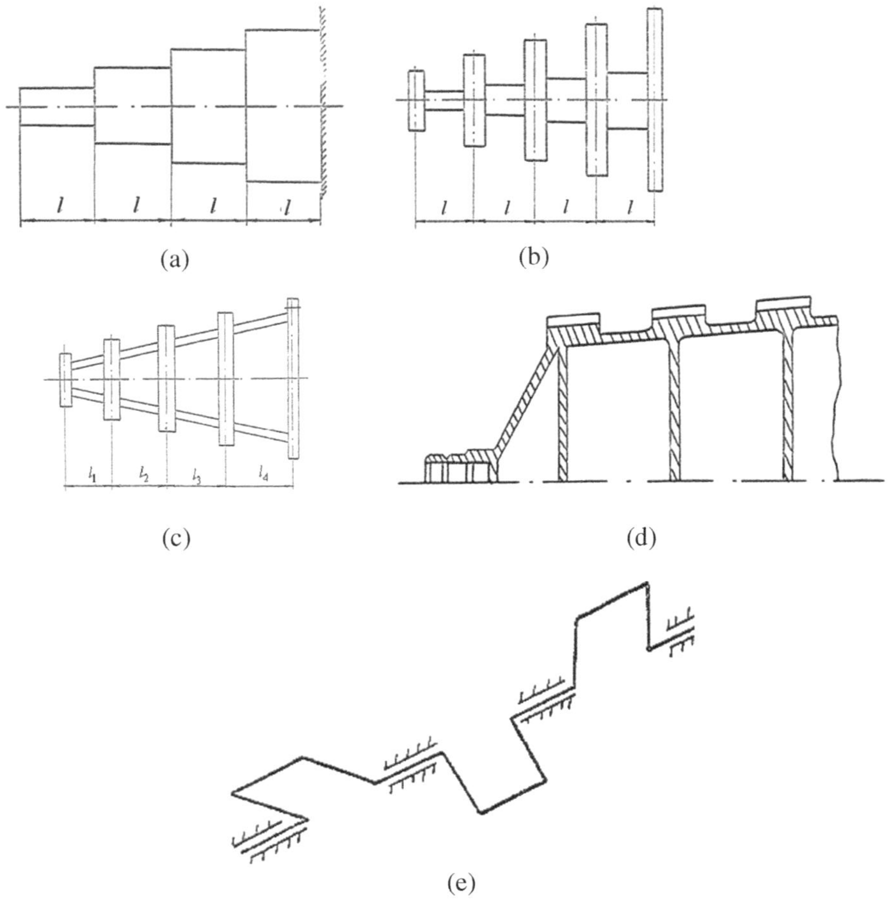

Fig. 9.1a-e Examples of self-similar structures: (**a**) step-up rod, (**b**) shaft whose parameters change stepwise by section, (**c**) conical shell, (**d**) section of a drum-type compressor rotor, (**e**) spatial crankshaft

9.2 Dynamic Compliances of Self-Similar Systems

Let us consider two elements of a system, k and $k+1$. Let us write for them the finite-difference equation (4.24)

$$-\mathbf{F}_k \mathbf{Q}[k-1] + (\mathbf{G}_{k+1} + \mathbf{T}_k)\,\mathbf{Q}[k] - \mathbf{\Phi}_{k+1}\mathbf{Q}[k+1] = 0. \tag{9.2}$$

It follows from Eq. (9.1) that

$$\begin{aligned} \mathbf{G}_{k+1} &= \gamma \mathbf{G}_k, \\ \mathbf{\Phi}_{k+1} &= \gamma \mathbf{\Phi}_k. \end{aligned} \tag{9.3}$$

Substituting (9.3) into (9.2) we obtain

$$-\mathbf{F}_k\mathbf{Q}\,[k-1]+(\gamma\mathbf{G}_k+\mathbf{T}_k)\,\mathbf{Q}\,[k]-\gamma\mathbf{\Phi}_k\mathbf{Q}\,[k+1]=0. \tag{9.4}$$

Matrices $\mathbf{\Phi}_k$ and $\mathbf{F}_k$ are transposed with respect to each other. Let $\mathbf{F}_1$, $\mathbf{G}_1$, $\mathbf{T}_1$, $\mathbf{\Phi}_1$ be the dynamic compliance matrices of the element $k=1$. Then, from Eq. (9.3) instead of Eq. (9.4)), we will have:

$$-\mathbf{F}_1\mathbf{Q}\,[k-1]+(\gamma\mathbf{G}_1+\mathbf{T}_1)\,\mathbf{Q}\,[k]-\gamma\mathbf{\Phi}_1\mathbf{Q}\,[k+1]=0. \tag{9.5}$$

This equation is a finite-difference matrix equation of the second order with constant coefficients. We will write its particular solution in the form

$$\mathrm{Q}\,[k]=\mathrm{A}CZ^k, \tag{9.6}$$

where $\mathbf{A}$ is an m-dimensional vector column and $\mathbf{C}$ is an arbitrary constant.

From Eqs. (9.5) and (9.6) we obtain

$$\left[\mathbf{F}_1\mathbf{Z}^{-1}+(\gamma\mathbf{G}_1+\mathbf{T}_1)+\gamma\mathbf{\Phi}_1\mathbf{Z}\right]\mathbf{A}=0. \tag{9.7}$$

Vector $\mathbf{A}$ is not equal to zero only if

$$\det\left[\mathbf{F}_1\mathbf{Z}^{-1}-(\gamma\mathbf{G}_1+\mathbf{T}_1)+\gamma\mathbf{\Phi}_1\mathbf{Z}\right]=0. \tag{9.8}$$

From Eq. (9.8) one can determine $2m$ values of $Z_r\,(r=1,2,\ldots,2m)$. Let us consider the case where all roots are different. We will write the general solution of Eq. (9.5) in the following form:

$$Q_r\,[k]=\sum_{s=1}^{2m}A_s^{(r)}C_sZ_s^k(r=1,2,\ldots,m). \tag{9.9}$$

The components $A_s^{(r)}$ of vector $\mathbf{A}_\mathrm{s}$ correspond to the value $Z_s\,(s=1,2,\ldots,2m)$; the arbitrary constants $\mathrm{C_s}$ are determined from the boundary conditions.

Let $\tilde{\mathbf{G}},\tilde{\mathbf{T}},\tilde{\mathbf{F}},\tilde{\mathbf{\Phi}}$ be, respectively, the input, output, direct, and inverse compliances matrices of the assembled system, $\mathbf{q}[0]$ and $\mathbf{q}[\mathrm{n}]$ the vector columns of the displacement amplitudes at the input and output of the system.

Then

$$\begin{aligned}\mathbf{q}\,[0]&=-\tilde{\mathbf{G}}\mathbf{Q}\,[0]+\tilde{\mathbf{\Phi}}\mathbf{Q}[n],\\ \mathbf{q}\,[n]&=-\tilde{\mathbf{F}}\mathbf{Q}\,[0]+\tilde{\mathbf{T}}\mathbf{Q}[n].\end{aligned} \tag{9.10}$$

On the other hand:

$$\begin{aligned}\mathbf{q}[0] &= -\mathbf{D}_1\mathbf{Q}[0] + \mathbf{\Phi}_1\mathbf{Q}[1] \\ \mathbf{q}[n] &= -\gamma^{n-1}\mathbf{F}_1\mathbf{Q}[n-1] + \gamma^{n-1}\mathbf{T}_1\mathbf{Q}[n].\end{aligned} \tag{9.11}$$

The dynamic compliances matrices for the whole system are determined from Eqs. (9.10) and (9.11) by taking into account Equality (9.9).

In the special case of a single-connectedness (m=1) self-similar system, the following analytical expressions for the dynamic compliances can be obtained:

$$\tilde{G} = G_1 - F_1\gamma^{-\frac{1}{2}}\frac{\sin(n-1)\varphi}{\sin n\varphi},$$

$$\tilde{F} = \tilde{\Phi} = F_1\gamma^{\frac{n-1}{2}}\frac{\sin\varphi}{\sin n\varphi} \tag{9.12}$$

$$\tilde{T} = T_1\gamma^{n-1} - F_1\gamma^{n-\frac{1}{2}}\frac{\sin(n-1)\varphi}{\sin n\varphi}$$

$$\cos\varphi = \frac{\gamma^{\frac{1}{2}}D_1 + \gamma^{-\frac{1}{2}}T_1}{2F_1}. \tag{9.13}$$

9.3 Vibrations of Self-Similar Systems: Numerical Examples

Let us now consider the free longitudinal vibrations of a step-up rod whose right-hand end is rigidly fixed and left-hand one is free (Fig. 9.1a), $\gamma = \frac{S_k}{S_{k+1}}$, where S_k is the area of the cross-section of the kth section.

The dynamic compliances of the first element are

$$G_1 = T_1 = F_1\cos\lambda,\ F_1 = \frac{l}{\lambda E' S_1 \sin\lambda},\ \lambda = \omega l\sqrt{\frac{\rho}{E'}}, \tag{9.13a}$$

where ρ is the mass density of the material, E is the elasticity modulus, and ω is the circular frequency.

When the right-hand end is fixed, $\tilde{T} = 0$, and since $F_1 \neq 0$, the frequency equation has the form

$$\cos\lambda = \frac{\sqrt{\gamma}\sin\left[(n-1)\arccos\frac{1+\gamma}{2\sqrt{\gamma}}\cos\lambda\right]}{\sin\left[n\arccos\frac{1+\gamma}{2\sqrt{\gamma}}\cos\lambda\right]}.$$

$$\text{At } n = 2,\ tg^2\lambda = \gamma = \frac{S_2}{S_1}.$$

9.4 Transition Matrix

For self-similar systems, according to Equality (9.1), we will have

$$\begin{aligned} \mathbf{G}_{k+1} &= \gamma \mathbf{G}_k, \boldsymbol{\Phi}_{k+1} = \gamma \boldsymbol{\Phi}_k, \\ \mathbf{F}_{k+1} &= \gamma \mathbf{F}_k, \mathbf{T}_{k+1} = \gamma \mathbf{T}_k. \end{aligned} \tag{9.14}$$

Using the relationships between the blocks of the transition matrix of the $(k+1)$th element through the corresponding blocks of the kth element, (see (4.17a).

$$\begin{aligned} \mathbf{W}_{11,k+1} &= \mathbf{W}_{11,k}, \mathbf{W}_{12,k+1} = \mathbf{W}_{12k}, \\ \mathbf{W}_{21,k+1} &= \frac{\mathbf{W}_{21k}}{\gamma}, \mathbf{W}_{22,k+1} = \mathbf{W}_{22k} \end{aligned} \tag{9.15}$$

Then we obtain for the transition matrix:

$$\mathbf{W}_{k+1} = \boldsymbol{\xi} \mathbf{W}_k \boldsymbol{\xi}^{-1}, \tag{9.16}$$

where

$$\boldsymbol{\xi} = \begin{bmatrix} \gamma \mathbf{E} & 0 \\ 0 & \mathbf{E} \end{bmatrix} \tag{9.17}$$

is a diagonal matrix ($2m \times 2m$) and $\mathbf{E}$ a identity matrix .

As is known (Chap. 5), the transition matrix for an assembled system is equal to the product of the transition matrices of its sections

$$\mathbf{W}^{(n)} = \prod_{k=n}^{1} \mathbf{W}_k = \mathbf{W}_n \mathbf{W}_{n-1} \ldots \mathbf{W}_1.$$

Let us set $\mathbf{W}_1 = \mathbf{W}$ for the first section; then

$$\begin{aligned} \mathbf{W}_2 &= \boldsymbol{\xi} \mathbf{W} \boldsymbol{\xi}^{-1} \\ \mathbf{W}_3 &= \boldsymbol{\xi}^2 \mathbf{W} \boldsymbol{\xi}^{-2} \\ \mathbf{W}_n &= \boldsymbol{\xi}^{(n-1)} \mathbf{W} \boldsymbol{\xi}^{-(n-1)}. \end{aligned}$$

And their product

$$\mathbf{W}^{(n)} = \gamma^n \left[\gamma^{-1} \mathbf{W} \right]^n. \tag{9.18}$$

Taking into account that

$$\boldsymbol{\xi}^n = \begin{bmatrix} \gamma^n \mathbf{E} & 0 \\ 0 & \mathbf{E} \end{bmatrix}$$

we will have for the blocks of matrix $\mathbf{W}^{(n)}$:

$$\mathbf{W}_{11}^{(n)} = \gamma^n \left[\boldsymbol{\xi}^{-1}\mathbf{W}\right]_{11}^n; \mathbf{W}_{12}^{(n)} = \gamma^n \left[\boldsymbol{\xi}^{-1}\mathbf{W}\right]_{12}^n;$$
$$\mathbf{W}_{21}^{(n)} = \left[\boldsymbol{\xi}^{-1}\mathbf{W}\right]_{21}^n; \mathbf{W}_{22}^{(n)} = \left[\boldsymbol{\xi}^{-1}\mathbf{W}\right]_{22}^n. \tag{9.19}$$

For the special case of a single-connectedness system, the transition matrix is of the second order. Its elements are:

$$w_{11}^{(n)} = w_{11}\gamma^{\frac{n-1}{2}}\frac{\sin n\varphi}{\sin\varphi} - \gamma^{\frac{n}{2}}\frac{\sin(n-1)\varphi}{\sin\varphi};$$
$$w_{21}^{(n)} = w_{21}\gamma^{-\frac{n-1}{2}}\frac{\sin n\varphi}{\sin\varphi};$$
$$w_{22}^{(n)} = w_{22}\gamma^{-\frac{n-1}{2}}\frac{\sin n\varphi}{\sin\varphi} - \gamma^{-\frac{n}{2}}\frac{\sin(n-1)\varphi}{\sin\varphi};$$
$$w_{11}^{(n)}w_{22}^{(n)} - w_{12}^{(n)}w_{21}^{(n)} = 1;$$
$$\cos\varphi = \frac{\gamma^{-\frac{1}{2}}w_{11} + \gamma^{\frac{1}{2}}w_{22}}{2}$$

where $w_{rs}(r, s = 1, 2)$ correspond to the first element of the system.

Let us now go back to the dependence between the natural frequencies and the normal vectors for the transition matrices of the two preceding elements.

Let us determine using Formula (6.20) the characteristic values of matrix $\mathbf{W}_\mathrm{k}$ and the corresponding normal vectors. Then matrix $\mathbf{W}_\mathrm{k}$ can be presented in the form as a product of three matrices [18]

$$\mathbf{W_k} = \mathbf{U_k}\boldsymbol{\Lambda}_\mathbf{k}\mathbf{U_k^{-1}}, \tag{9.20}$$

where $\mathrm{U_k}$ is a matrix $(2m\times 2m)$ consisting of the normal vectors and Λ_k is a diagonal matrix $(2m\times 2m)$ of the characteristic values.

Let us write (9.20) in block form

$$\begin{bmatrix}\mathbf{W}_{11k} & \mathbf{W}_{12k}\\ \mathbf{W}_{21k} & \mathbf{W}_{22k}\end{bmatrix} = \begin{bmatrix}\mathbf{U}_{11k} & \mathbf{U}_{12k}\\ \mathbf{U}_{21k} & \mathbf{U}_{22k}\end{bmatrix}\begin{bmatrix}\boldsymbol{\Lambda}_{1k} & 0\\ 0 & \boldsymbol{\Lambda}_{2k}\end{bmatrix}\begin{bmatrix}\mathbf{U}_{11k}^* & \mathbf{U}_{12k}^*\\ \mathbf{U}_{21k}^* & \mathbf{U}_{22k}^*\end{bmatrix}. \tag{9.21}$$

Every block in this equation is a square matrix $(m\times m)$, $\boldsymbol{\Lambda}_{1k}$ and $\boldsymbol{\Lambda}_{2k}$ are diagonal matrices, and $\mathbf{U}_{11k}^*$, $\mathbf{U}_{12k}^*$, $\mathbf{U}_{21k}^*$, $\mathbf{U}_{22k}^*$ are the blocks of the inverse matrix $\mathbf{U}_k^{-1}$.

For each element k+1 from Eq. (9.16) we have

$$\mathbf{W}_{k+1} = \begin{bmatrix}\gamma\mathbf{E} & 0\\ 0 & \mathbf{E}\end{bmatrix}\begin{bmatrix}\mathbf{W}_{11k} & \mathbf{W}_{12k}\\ \mathbf{W}_{21k} & \mathbf{W}_{22k}\end{bmatrix}\begin{bmatrix}\gamma^{-1}\mathbf{E} & 0\\ 0 & \mathbf{E}\end{bmatrix}. \tag{9.22}$$

Substituting Eq. (9.21) into (9.22) we find

$$\mathbf{W}_{k+1} = \begin{bmatrix}\gamma\mathbf{E} & 0\\ 0 & \mathbf{E}\end{bmatrix}\begin{bmatrix}\mathbf{U}_{11k} & \mathbf{U}_{12k}\\ \mathbf{U}_{21k} & \mathbf{U}_{22k}\end{bmatrix}\begin{bmatrix}\boldsymbol{\Lambda}_{1k} & 0\\ 0 & \boldsymbol{\Lambda}_{2k}\end{bmatrix}\begin{bmatrix}\mathbf{U}_{11k}^* & \mathbf{U}_{12k}^*\\ \mathbf{U}_{21k}^* & \mathbf{U}_{22k}^*\end{bmatrix}\begin{bmatrix}\gamma^{-1}\mathbf{E} & 0\\ 0 & \mathbf{E}\end{bmatrix}. \tag{9.23}$$

On the other hand, by writing (9.21) for matrix $\mathbf{W}_{k+1}$, we obtain

$$\mathbf{W}_{k+1} = \begin{bmatrix} \mathbf{U}_{11\,k+1} & \mathbf{U}_{12\,k+1} \\ \mathbf{U}_{21\,k+1} & \mathbf{U}_{22\,k+1} \end{bmatrix} \begin{bmatrix} \boldsymbol{\Lambda}_{1k}+1 & 0 \\ 0 & \boldsymbol{\Lambda}_{2k}+1 \end{bmatrix} \begin{bmatrix} \mathbf{U}^*_{11\,k+1} & \mathbf{U}^*_{12\,k+1} \\ \mathbf{U}^*_{21\,k+1} & \mathbf{U}^*_{22\,k+1} \end{bmatrix}. \tag{9.24}$$

It follows from Shur's formula that the characteristic values of matrices $\mathbf{W}_k$ and $\mathbf{W}_{k+1}$, whose blocks are related by relationship (9.23), are equal to each other.
Then

$$\boldsymbol{\Lambda}_{1,k+1} = \boldsymbol{\Lambda}_{1k}, \boldsymbol{\Lambda}_{2,k+1} = \boldsymbol{\Lambda}_{2k}. \tag{9.25}$$

The dependencies between the normal vectors are determined by comparing the left-hand parts of Equalities (9.23) and (9.24)

$$\mathbf{U}_{11,k+1} = \gamma\mathbf{U}_{11\,k},\ \mathbf{U}_{12,k+1} = \gamma\mathbf{U}_{12\,k},\ \mathbf{U}_{21,k+1} = \gamma\mathbf{U}_{21\,k},\ \mathbf{U}_{22,k+1} = \gamma\mathbf{U}_{22\,k}. \tag{9.26}$$

It can be seen from this that m components of the normal vectors of the transition matrices $\mathbf{W}_k$ and $\mathbf{W}_{k+1}$ are similar (coefficient of similarity γ), and the remaining m components are equal to each other.

9.5 Self-Similar Systems with Similar Matrix of Dynamic Compliance

So far we have been discussing self-similar systems in which condition (9.1) between the dynamic compliances matrices of two neighboring elements was kept. The more general case of a self-similar system, where the value γ is different for the forced, mixed, and moment dynamic compliances, can be treated analogously. In particular, some types of rotors with a drum structure belong to systems of this type.

Let us write in block form the matrix equation (4.24) for the self-similar system under consideration

$$-\begin{bmatrix} \mathbf{F}_{11\,k} & \mathbf{F}_{12\,k} \\ \mathbf{F}_{21\,k} & \mathbf{F}_{22\,k} \end{bmatrix} \begin{bmatrix} \mathbf{Q}_{1,\mathrm{k}\text{-}1} \\ \mathbf{Q}_{2,\mathrm{k}\text{-}1} \end{bmatrix} + \begin{bmatrix} \mathbf{G}_{11,k+1} + \mathbf{T}_{11\,k} & \mathbf{G}_{12,k+1} + \mathbf{T}_{12\,k} \\ \mathbf{G}_{21,k+1} + \mathbf{T}_{21\,k} & \mathbf{G}_{22,k+1} + \mathbf{T}_{22\,k} \end{bmatrix} \\ \begin{bmatrix} \mathbf{Q}_{1\mathrm{k}} \\ \mathbf{Q}_{2\mathrm{k}} \end{bmatrix} - \begin{bmatrix} \boldsymbol{\Phi}_{11,k+1} & \boldsymbol{\Phi}_{12,k+1} \\ \boldsymbol{\Phi}_{21,k+1} & \boldsymbol{\Phi}_{22,k+1} \end{bmatrix} \begin{bmatrix} \mathbf{Q}_{1,\mathrm{k}+1} \\ \mathbf{Q}_{2,\mathrm{k}+1} \end{bmatrix} = 0. \tag{9.27}$$

The index 11 designates the dynamic compliance from forces , the index 22 from the moments one, the indices 12 and 21 the mixed one. Vector $\mathbf{Q}_1$ corresponds to the translational displacements, $\mathbf{Q}_2$ to the angular ones.

Let the following relationships be valid for the dynamic compliances of the k-th and the (k+1)-th elements:

$$\mathbf{L}_{11,k+1} = \gamma \mathbf{L}_{11\,k}, \mathbf{L}_{12,k+1} = \gamma^2 \mathbf{L}_{12\,k}, \mathbf{L}_{21,k+1} = \gamma^2 \mathbf{L}_{21\,k},$$

$$\mathbf{L}_{22,k+1} = \gamma^3 \mathbf{L}_{22\,k}, \;\; \mathbf{L} = \mathbf{F,G,T,\Phi}. \tag{9.27a}$$

From Eq. (9.27):

$$\begin{bmatrix} \mathbf{F}_{11} & \mathbf{F}_{12}\gamma^{k-1} \\ \mathbf{F}_{21} & \mathbf{F}_{22}\gamma^{k-1} \end{bmatrix} \begin{bmatrix} \mathbf{Q}_1\,[k-1] \\ \mathbf{Q}_2\,[k-1] \end{bmatrix} - \begin{bmatrix} \mathbf{T}_{11} + \gamma \mathbf{G}_{11} & \mathbf{T}_{12}\gamma^{k-1} + \mathbf{G}_{12}\gamma^{k+1} \\ \mathbf{T}_{21} + \gamma^2 \mathbf{G}_{21} & \mathbf{T}_{22}\gamma^{k-1} + \mathbf{G}_{22}\gamma^{k+2} \end{bmatrix} \begin{bmatrix} \mathbf{Q}_1\,[k] \\ \mathbf{Q}_2\,[k] \end{bmatrix} +$$
$$+ \begin{bmatrix} \mathbf{\Phi}_{11}\gamma & \mathbf{\Phi}_{12}\gamma^{k+1} \\ \mathbf{\Phi}_{21}\gamma^2 & \mathbf{\Phi}_{22}\gamma^{k+2} \end{bmatrix} \begin{bmatrix} \mathbf{Q}_1\,[k+1] \\ \mathbf{Q}_2\,[k+1] \end{bmatrix} = 0. \tag{9.28}$$

This equation is a finite-difference matrix equation with coefficients depending from γ. Its particular solution is

$$\begin{aligned} \mathrm{Q}_1 &= \mathrm{A}_1 C Z^k, \\ \mathrm{Q}_2 &= \mathrm{A}_2 C \varepsilon^{-k} Z^k. \end{aligned} \tag{9.29}$$

From Eq. (9.28) it can be found that:

$$\begin{bmatrix} \mathbf{b}_{11} & \mathbf{b}_{12} \\ \mathbf{b}_{21} & \mathbf{b}_{22} \end{bmatrix} \begin{bmatrix} \mathrm{A}_1 \\ \mathrm{A}_2 \end{bmatrix} = 0, \tag{9.30}$$

here $\mathbf{b}_{ij}$ are the matrix depending from **F,G,T,Φ**

Then the characteristic equation will be

$$\det \begin{bmatrix} \mathbf{b}_{11} & \mathbf{b}_{22} \\ \mathbf{b}_{21} & \mathbf{b}_{22} \end{bmatrix} = 0. \tag{9.31}$$

From here we determine the $2m$ values of $\mathrm{Z_r}(\mathrm{r} = 1, 2, \ldots, 2\mathrm{m})$.

If all $\mathrm{Z_r}$ are different, the general solution is written as follows:

$$\begin{aligned} Q_{1_1}^{(r)}\,[k] &= \sum_{s=1}^{l} A_s^r C_s Z_s^k (r = 1, 2, \ldots, l), \\ Q_2^{(t)}\,[k] &= \sum_{s=1}^{f} A_s^t C_s \gamma^{-k} Z_s^k (t = 1, 2, \ldots, f). \end{aligned} \tag{9.32}$$

The arbitrary constants $\mathrm{C_s}$ are determined from the boundary conditions.

9.6 Vibrations of Self-Similar Shaft with Disks

The view of such shaft is shown in Fig. 9.2a. The disks are connected to each other by means of inertia-free connections – beams. The homothety coefficient of the system is γ. This means that the following relationships are true:

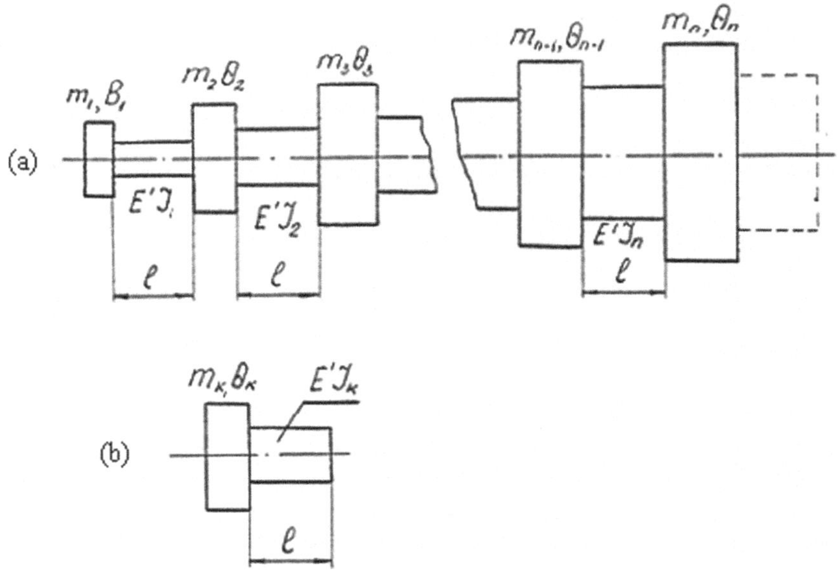

Fig. 9.2 A self-similar shaft with disks (**a**) and a section of the shaft (**b**)

$$m_k = \gamma m_{k+1},\ \theta_k = \gamma\theta_{k+1},\ E'J_k = \gamma E'J_{k+1},\ l_{k+1} = l_k = l,\ \gamma < 0,$$

where m_k and θ_k are the mass and the axial inertia moment of the kth disk, $E'J$ is the bending rigidity of the beam, and l is the distance between the disks.

The dependence between the vectors at the input and the output of the kth element has the form

$$\mathbf{X}_{kb} = \mathbf{W}\mathbf{X}_{k\alpha},$$

where

$$\mathbf{X}_{ks} = \left[y_s, \varphi_s, \tilde{M}_s, Q_s\right] \quad (s = a, b). \tag{9.33}$$

Where y_s, φ_s are the displacement and the rotation angle of the section, $Q_s, \tilde{M}_s$ are the transversal force and the bending moment, and $\mathbf{W}_b$, $\mathbf{W}_d$ are the transition matrices for the beam and the disk, respectively. Then $\mathbf{W} = \mathbf{W}_b\mathbf{W}_d$ is the transition matrix for the section of the beam in Fig. 9.2.

Let us determine the transition matrix $\mathbf{W}_1$ for the first section. The transition matrix of the inertia-free section is determined from the Eq. (7.17) where, instead of $E'J_b$, must be used $E'J_1$ (where J_1 is the axial inertia moment of the shaft cross section in the first section), and l must be used instead of t_n. The transition matrix for the disk is equal to

$$\mathbf{W}_d = \begin{bmatrix} 1 & 0 & 0\,0 \\ 0 & 1 & 0\,0 \\ 0 & -\theta_1\omega^2 & 1\,0 \\ -m_1\omega^2 & 0 & 0\,1 \end{bmatrix}.$$

After multiplying the matrices, we obtain for $\mathbf{W}_1$

$$\mathbf{W}_1 = \begin{bmatrix} 1 & l & \frac{l\beta_1}{2} & -\frac{l^2\beta_1}{6} \\ 0 & 1 & \beta_1 & -\frac{l\beta_1}{2} \\ 0 & -\theta_1\omega^2 & 1-\theta_1\omega^2 & -l+\theta_1\omega^2\frac{l\beta_1}{2} \\ -m_1\omega^2 & -m_1\omega^2 l & -m_1\omega^2\frac{l\beta_1}{2} & 1+m_1\omega^2\frac{l\beta_1}{6} \end{bmatrix}, \; \beta_1 = \frac{l}{E'J_1}.$$

With matrix $\mathbf{W}_1$ it is possible by using Eqs. (9.18) and (9.19) to determine the transition matrix for the whole shaft. Next, the natural frequencies and amplitudes of the forced vibrations are determined. We would like to point out that by using the transition matrix the shaft under consideration can be included as a subsystem in a more complex nonregular system.

9.7 Vibrations of Self-Similar Drum-Type Rotor

A model of this rotor is shown in Fig. 9.3a, b. The disks are linked to each other with inertia-free connections. These connections are conical shells with variable thickness.

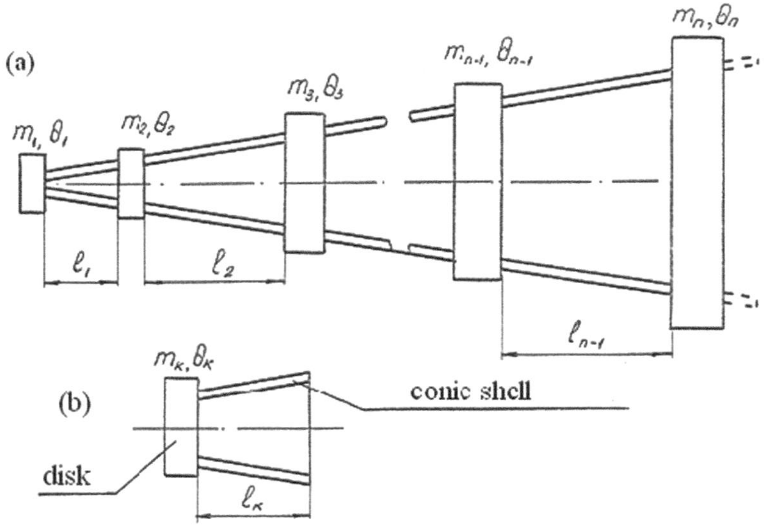

Fig. 9.3 A self-similar rotor of the drum type (**a**) and a section of the rotor (**b**)

Let us consider a standard element of the system (Fig. 9.3b). This element consists of two elements: a weightless conical shell and a disk with mass m and axial inertia moment θ. Let us write the dependence between the forces and the displacements at the input and the output of the element

$$\begin{bmatrix} y_a \\ \varphi_a \\ y_b \\ \varphi_b \end{bmatrix} = \begin{bmatrix} G_{11} & G_{12} & F_{11} & F_{12} \\ G_{21} & G_{22} & F_{21} & F_{22} \\ \Phi_{11} & \Phi_{12} & T_{11} & T_{12} \\ \Phi_{21} & \Phi_{22} & T_{21} & T_{22} \end{bmatrix} \begin{bmatrix} -Q_a \\ -M_a \\ -Q_b \\ -M_b \end{bmatrix}. \tag{9.34}$$

where $y_a, \varphi_a (y_b, \varphi_b)$ are, respectively, the displacement and the rotation angle at the input (output) of the element; $\mathrm{Q_a}, \mathrm{M_a}(\mathrm{Q}_\alpha, \mathrm{M}_\alpha)$ are, respectively, the transversal force and the bending moment at the input (output) of the element.

In Eqs. (9.34), it was assumed that the shells were sufficiently long, so that the effect of the self-balanced end loads acting at the input (output) of a shell on its displacements at the input (output) [106] could be neglected.

$$\begin{aligned} &G_{11} = F_{11} = \Phi_{11} = -\frac{1}{m\omega^2},\ G_{22} = F_{22} = \Phi_{22} = -\frac{1}{\theta\omega^2}, \\ &G_{12} = G_{21} = F_{12} = \Phi_{21} = 0,\ F_{21} = \Phi_{12} = -\frac{l}{\theta\omega^2}, \\ &T_{11} = e_{11\alpha} - \frac{1}{m\omega^2} - \frac{l}{\theta\omega^2},\ T_{21} = T_{12} = e_{12\alpha} - \frac{l}{\theta\omega^2}, \\ &T_{22} = e_{22\alpha} - \frac{l}{\theta\omega^2} \end{aligned} \tag{9.35}$$

where $e_{11b}, e_{12b} = e_{21b}, e_{22b}$ are the forced, the mixed, and the momentous dynamic compliance at the output of the conical shell under the condition that the input section has been rigidly fixed (Fig. 9.4a).

Let us now write the dependencies between the output and input dynamic compliances of the conical shell. The input compliances $e_{12\alpha} = e_{21\alpha}, e_{22\alpha}$ are calculated under the condition that the shell is rigidly fixed at the output. The dependence between the input and output compliances can be found with the help of Eqs. (A1.3) and (A1.6) (Appendix A). Since the conical shell is a two-connectedness system, all block matrices in these formulas turn into scalars, Then

$$\boldsymbol{\theta} = \boldsymbol{\theta}^T = 1, v_x = l.$$

$$\begin{aligned} &e_{11b} = e_{11a} + 2le_{21a} + l^2 e_{22a}, \\ &e_{12b} = e_{21\alpha} = e_{12a} + le_{22a},\ \ e_{22\alpha} = e_{22a}. \end{aligned} \tag{9.36}$$

From (9.13a), (9.35) [109]:

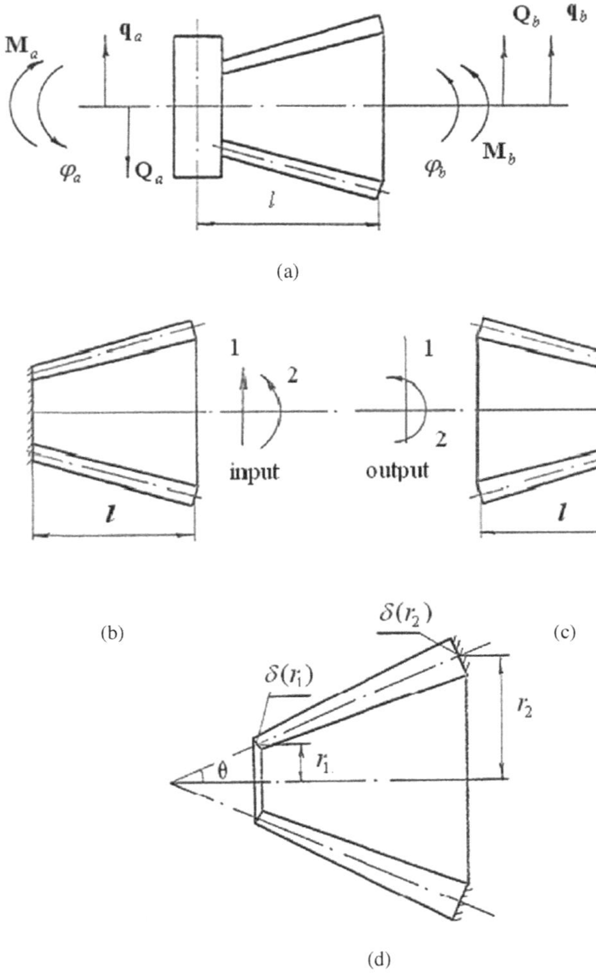

Fig. 9.4 Element of a self-similar system: disk and a conical shell with variable thickness

$$
\begin{aligned}
e_{11a} &= \frac{1}{E'\delta\left(r_1\right)\sin\gamma}e_1, \; e_{12a} = e_{21a} = \frac{1}{E'r_1\delta\left(r_1\right)\cos\gamma}e_2, \\
e_{22a} &= \frac{1}{E'r_1^2\delta\left(r_1\right)},
\end{aligned} \tag{9.37}
$$

where E' is the elasticity modulus, $\delta\left(r_1\right)$ is the thickness of the shell, and r_1 is the radius of the middle surface of the shell at the input,

$$
e_i = f_i\left(\beta,\gamma,\xi\right) \;\; (i = 1, 2, 3), \; \beta = \frac{r_1}{r_2}, \; \xi = \frac{\delta\left(r_2\right)}{\delta\left(r_1\right)}.
$$

The remaining notations are given in Fig. 9.4.

Let us assume that the following relationships are valid for two rotor elements with numbers k and $k+1$:

$$\begin{aligned} &m_k = \gamma m_{k+1},\ \theta_k = \gamma^3\theta_{k+1},\ r_{1k} = \gamma r_{1,k+1},\ \delta\left(r_{1k}\right) = \gamma\delta\left(r_{1,k+1}\right),\\ &l_k = \gamma l_{k+1},\ \xi_k = \xi_{k+1},\ \theta_k = \theta_{k+1},\ \beta_k = \beta_{k+1}. \end{aligned} \tag{9.38}$$

Then from Eqs. (9.35), (9.36), and (9.37):

$$\begin{aligned} &G_{11,k+1} = \gamma G_{11\,k};\ F_{11,k+1} = \gamma F_{11\,k};\ \Phi_{11,k+1} = \gamma\Phi_{11\,k};\\ &T_{11,k+1} = \gamma T_{11\,k};\ F_{21,k+1} = \gamma^2 F_{21\,k};\ \Phi_{12,k+1} = \gamma^2\Phi_{12\,k};\\ &T_{12,k+1} = \gamma^2 T_{12\,k};\ T_{21,k+1} = \gamma^2 T_{21\,k};\ G_{22,k+1} = \gamma^3 G_{22\,k};\\ &F_{22,k+1} = \gamma^3 D_{22\,k};\ \Phi_{22,k+1} = \gamma^3\Phi_{22\,k};\ T_{22,k+1} = \gamma^3 T_{22\,k}. \end{aligned} \tag{9.39}$$

Relationships (9.39) are equivalent to (9.28). Therefore, the drum-type rotor can be treated as a self-similar system with similar blocks of dynamic compliance matrix. The calculation of such a rotor can be performed with Eqs. (9.29), (9.30), (9.32), (9.32), (9.33), and (9.34).

Chapter 10
Vibrations of Rotor Systems with Periodic Structure

The problem of the existence of critical velocities at the forward and backward rotor precession in the absence of friction forces is treated in [31,81]. As is known, solving such a problem consists in determining the natural frequencies of the bending vibrations of a no rotating shaft with disks whose equatorial inertia moments are determined from the following equations [30]:

– at forward precession

$$\theta^{(np)} = \theta_o - \frac{\Omega}{\omega}\theta_z, \tag{10.1}$$

– at backward precession

$$\theta^{(np)} = \theta_o + \frac{\Omega}{\omega}\theta_z, \tag{10.2}$$

where θ_o and θ_z are, respectively, the equatorial and axial inertia moments of the disk, Ω is the angular velocity of rotation, and ω is the vibrations frequency.

10.1 Rotor Systems with Periodic Structure with Disks

Let us set $\Omega = 0$ in Eqs. (10.1) and (10.2), i.e., we will calculate the dynamic characteristics of a nonrotating rotor. However, all results – as is known from [81, 106] – can be obtained also for a rotating rotor.

The calculation is based on the discrete model of the shaft – disks connected by inertia-free elements (beams). In this way, the beams work on bending. A calculation model of the shaft is shown in Fig. 10.1a. This is a two-connectedness system with reflection symmetry elements. The regular element of the system is presented in Fig. 10.1b. It consists of two disks connected to each other by means of a weightless beam. The mass of the disk is $\frac{m}{2}$; its axial moment of inertia is $\frac{\theta}{2}$. Rigid supports are mounted in the gravity center of the disks. We will determine the mixed dynamic matrix of the element. As vector $\mathbf{Z}(k)$ we select

L.Y. Banakh, M.L. Kempner, *Vibrations of Mechanical Systems with Regular Structure*, Foundations of Engineering Mechanics, DOI 10.1007/978-3-642-03126-7_10,

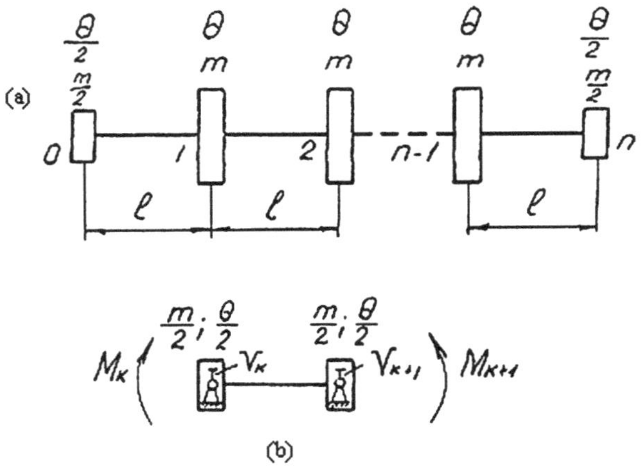

Fig. 10.1 Rotor with periodic structure (a) and an element of the rotor (b)

$$\mathbf{Z}(k) = \begin{bmatrix} M(k) \\ V(k) \end{bmatrix} \quad (k = 1, 2, \ldots, n - 1),$$

where $M(k)$ is the bending moment at the input of the element and $V(k)$ the vertical displacement of the disk's center of gravity.

We determine the elements of mixed matrices $\boldsymbol{P^*}$ and $\boldsymbol{N}$ by Eqs. (7.25) and (7.26)

$$G = \frac{l}{3E'J} \frac{1 - 3\beta}{(1 - 6\beta)(1 - 2\beta)},$$

$$G^* = \frac{l}{6E'J} \frac{1}{(1 - 6\beta)(1 - 2\beta)}, \quad S = -S^* = -\frac{1}{l} \frac{1}{1 - 2\beta},$$

$$R = -\frac{m\omega^2}{2} \left(1 + \frac{\tilde{J}}{ml^2} \frac{2}{1 - 2\beta} \right), \quad R^* = -\frac{\tilde{J}\omega^2}{l} \frac{1}{1 - 2\beta},$$

where m and $\tilde{J}$ are the mass and the inertia moment of the disk, $E'J$ is the rigidity of the beam to bending, l is the beam length, and $\beta = \frac{\tilde{J}l\omega^2}{24E'J}$.

Let us write Eq. (7.33) for determining the characteristic values of the mixed matrix in expanded form:

$$\det \begin{bmatrix} G - \lambda G^* & -S + \lambda S^* \\ -S + \lambda S^* & -R + \lambda R^* \end{bmatrix} = 0.$$

Expanding the determinant, we obtain the following equation:

$$\alpha (3 - z - 3\alpha\psi)(1 - \alpha\psi + 2\psi z) = z^2 (1 - 3\alpha\psi), \tag{10.3}$$

where

$$z = 1 - \lambda, \tag{10.4}$$

$$\alpha = \frac{2\beta}{\psi} = \frac{ml^3\omega^3}{12E'J}, \tag{10.5}$$

$$\psi = \frac{\theta}{ml^2}. \tag{10.6}$$

It follows from Eq. (10.3) that the characteristic values λ are functions of parameter α and coefficient ψ.

Let us solve Eq. (10.3) for z, and, taking into account (10.4), we obtain the following expression for λ:

$$\lambda_{1,2} = 1 - 3\alpha\psi = \frac{\alpha}{2} \pm \sqrt{\alpha\,[36\alpha\,(\psi - 0.622)\,(\psi - 0.04465) + 12}. \tag{10.7}$$

Three intervals for the change in ψ are of interest: (1) $\psi \leq 0.04465$; (2) $0.04465 < \psi < 0.6220$; (3) $\psi \geq 0.6220$.

At $\psi \leq 0.04465$, the both characteristic values $\lambda_1(\alpha)$ and $\lambda_2(\alpha)$ are real at all α. The character of the change in these functions ($\psi = 0.02$) is presented in Fig. 10.2. Beginning with some value of α, both characteristic values λ_1 and λ_2 always remain positive, whereby $\lambda_1(\alpha) > 1$.

At $\psi \geq 0.6220$, both characteristic values are also always real. Beginning with a certain value of α, both characteristic values $\lambda_1(\alpha)<0$ and $\lambda_2(\alpha) < 0$, whereby in the entire range $\lambda_2(\alpha) < 1$. Figure 10.3 shows the functions $\lambda_1(\alpha)$ and $\lambda_2(\alpha)$ at $\psi = 1$.

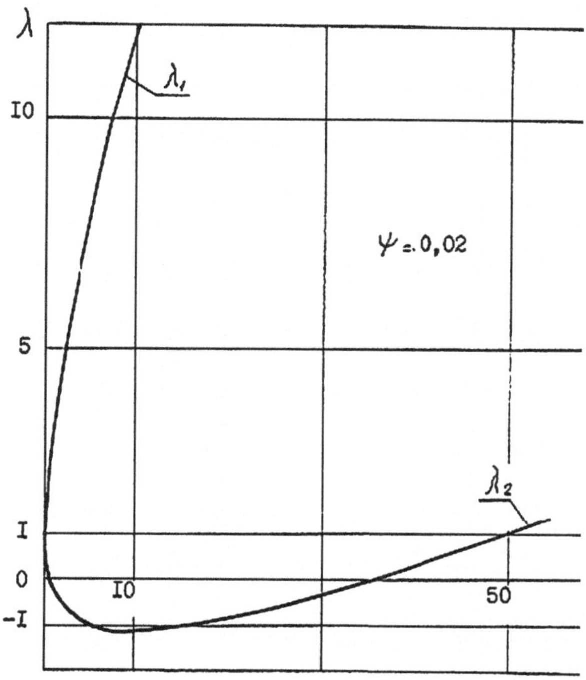

Fig. 10.2 Characteristic values of the mixed matrix for an element Fig. 10.1b

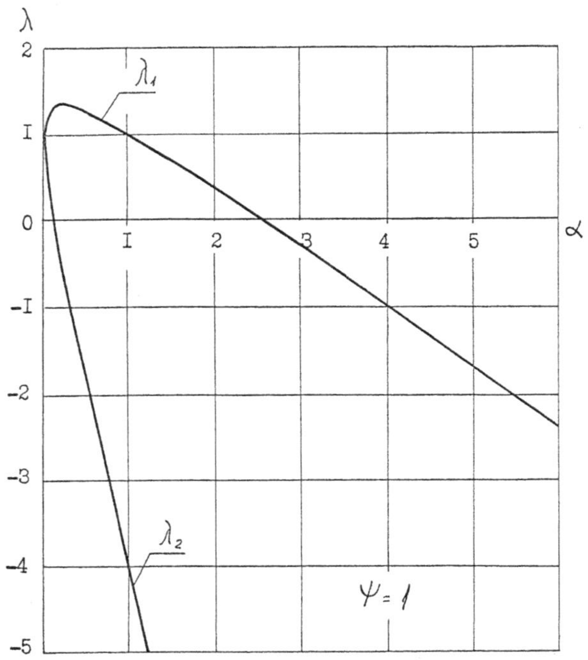

Fig. 10.3 Characteristic values of the mixed matrix of an element (Fig. 10.1b)

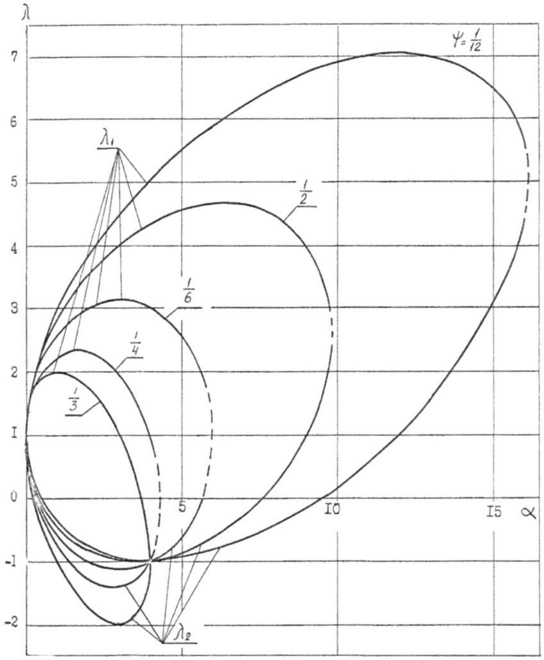

Fig. 10.4 Characteristic values of the mixed matrix of an element (Fig. 10.1b)

At $0.04465 < \psi < 0.6220$, the mixed matrices $\mathbf{P}^*$ and $\mathbf{N}$ are no longer positively (negatively) determined in the whole range of values of α. Therefore, the characteristic values can be obtained both as real and as complex-conjugated. The real values of λ_1 and λ_2 (solid lines) are shown in Fig. 10.4 at $\psi = \frac{1}{12}, \frac{1}{6}, \frac{1}{4}, \frac{1}{3}, \frac{1}{2}$. The minimum value $\alpha = 4$, at which λ_1 and λ_2 are still real, corresponds to the value $\psi = \frac{1}{3} = 0.333$. This is the value of ψ at which the radicand in Formula (10.7) has a minimum.

10.2 Rotor with Arbitrary Boundary Conditions: Natural Frequencies and Normal Modes

Using the results from Chap. 8, we can determine the frequencies and normal modes of vibration of a shaft at any boundary condition and also find the elements of the mixed matrix for the whole shaft.

Let us consider now the particular case of boundary conditions of the first type (Chap. 8) where the vertical displacements of the end disks are equal to zero and there is no bending moments at the ends of the system (Fig. 10.5). In order to obtain the natural frequencies of such a shaft, it is necessary to equalize

$$\lambda_{1,2} = \cos\frac{s\pi}{n}, \quad (s = 1, 2, \ldots, n-1), \tag{10.8}$$

n is the number of elements.

There are two trivial frequencies at which $M_k = 0$, $\alpha' = 1$, and $\alpha'' = 3$.

At $\alpha' = 1$, the normal mode of the shaft has the view shown in Fig. 10.6a. The regular elements of the shaft are self-balanced during vibrations and the support reactions at points 0 and n are equal to zero.

At $\alpha' = 3$, the normal mode of the shaft has the view shown in Fig. 10.6b. Here, as in the first case, there is no bending moments in the sections between the regular elements. However, the elements and the weight of the shaft are not self-balanced and reactions emerge in the supports during vibrations.

A shaft with four disks, number of regular elements $n = 3$, $\psi = \frac{1}{3}$ is considered as an example. From Eq. (10.8) we obtain four frequency values (Fig. 10.7).

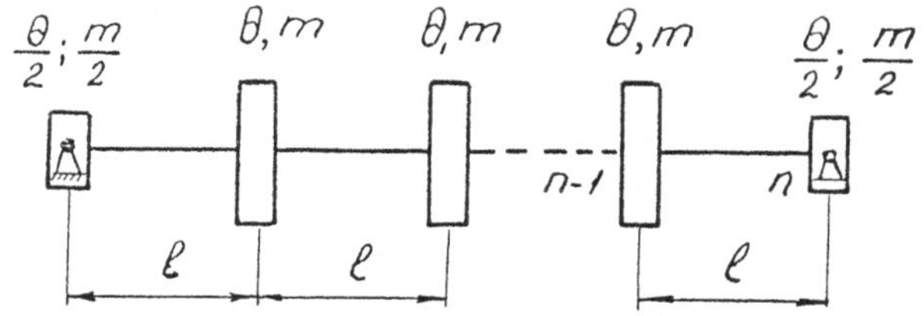

Fig. 10.5 Boundary conditions of the first type for a shaft with disks

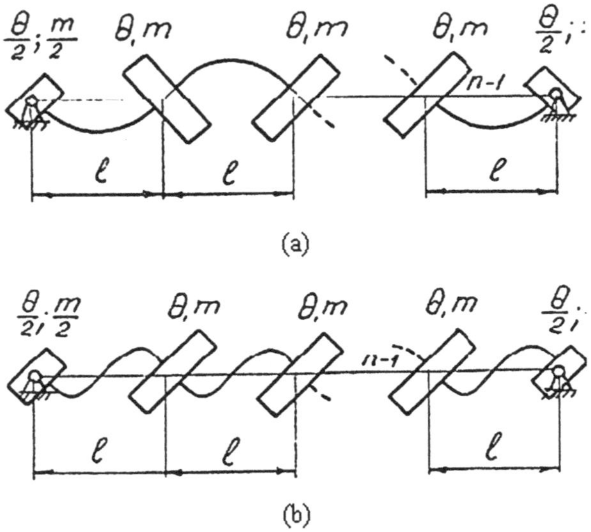

Fig. 10.6 Normal modes for a shaft with disks at $\alpha' = 1$ (**a**) and at $\alpha''' = 3$ (**b**)

$$\alpha_1 = 0.1;\ \alpha_2 = 0.7;\ \alpha_5 = 3.25;\ \alpha_6 = 3.95,$$

and two more trivial frequencies $\alpha_3 = 1,\ \alpha_4 = 3$.

Let us consider now the limit cases where the masses and the moments of inertia of the disks are equal to zero or tend to infinity.

a) $\theta = 0, \psi = 0$

$$\omega^2 = 12\frac{E'J}{ml^3}\frac{\left(1-\cos\frac{s\pi}{n}\right)^2}{2+\cos\frac{s\pi}{n}}, (s = 1, 2, \dots, n-1).$$

Quite naturally, the result coincides with the solution for a beam discussed in Sect. 8.5, example 4.

b) $\theta \to \infty, \psi \to \infty$

$$\omega^2 = \frac{24E'J}{ml^3}\left(1-\cos\frac{s\pi}{n}\right)(s = 1, 2, \dots, n-1)$$

c) $m \to 0, \psi \to \infty$

$$\omega^2 = \frac{2E'J}{\theta l}\left(1-\cos\frac{s\pi}{n}\right)(s = 1, 2, \dots, n)$$

d) $m \to \infty, \psi \to 0$

$$\omega^2 = \frac{2E'J}{\theta l}\left(1+\cos\frac{s\pi}{n}\right)(s = 1, 2, \dots, n).$$

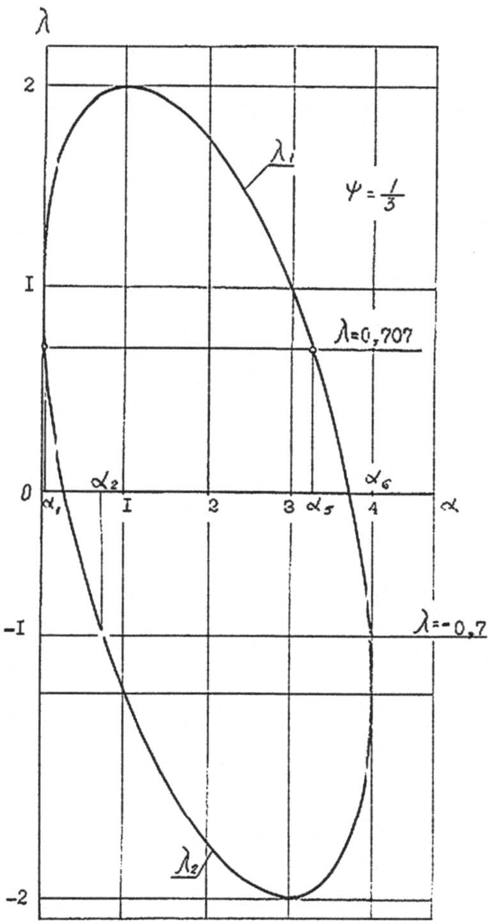

Fig. 10.7 Determining the natural frequencies of a shaft with disks at boundary conditions of the first type

Chapter 11
Vibrations of Regular Ribbed Cylindrical Shells

11.1 General Theory of Shells

This chapter is devoted to the study of oscillations of shells supported by transversal or longitudinal stiffening ribs. The relationships between the forces and the deformations of the middle surface of the shell proposed by V.V. Novozhilov and L.I. Balabukh [83] are used. The main advantage of these equations is that in them Betty's law is observed very precisely and thus the dynamic stiffness matrix is symmetric, and therefore it is possible to use all results that have been obtained above.

Let us consider the free vibrations of a circular cylindrical shell. Let us select the coordinate system and the direction of the displacements as shown in Fig. 11.1.

The differential equations of a shell for displacements have the form [83]

$$\begin{aligned}
&\frac{\partial^2 u}{\partial \xi^2} + \frac{1-\mu}{2}\frac{\partial^2 u}{\partial \varphi^2} + \frac{1+\mu}{2}\frac{\partial^2 v}{\partial \xi \partial \varphi} + \mu\frac{\partial \omega}{\partial \xi} - \frac{r^2 h\varsigma}{D}\frac{\partial^2 u}{\partial t^2} = 0\\
&\frac{1+\mu}{2}\frac{\partial^2 v}{\partial \xi \partial \varphi} + \frac{1-\mu}{2}(1-4\kappa)\frac{\partial^2 v}{\partial \xi^2} + (1+\kappa)\frac{\partial^2 v}{\partial \varphi^2} + \frac{\partial \omega}{\partial \varphi} -\\
&-\kappa\frac{\partial^3 w}{\partial \varphi^3} - \kappa(2-\mu)\frac{\partial^3 W}{\partial \xi^3 \partial \varphi^3} - \frac{r^2 h\rho}{D}\frac{\partial^2 v}{\partial t^2} = 0, \qquad (11.1)\\
&\mu\frac{\partial u}{\partial \xi} + \frac{\partial v}{\partial \varphi} - \kappa\frac{\partial^3 v}{\partial \varphi^3} - \kappa(2-\mu)\frac{\partial^3 v}{\partial \xi^2 \partial \varphi} + \omega +\\
&+\kappa\left(\frac{\partial^4 w}{\partial^4} + 2 - \frac{\partial^4 w}{\partial \xi^2 \partial \varphi^2} + \frac{\partial^4 w}{\partial \varphi^4}\right) + \frac{r^2 h\rho}{D}\frac{\partial^2 w}{\partial t^2} = 0,
\end{aligned}$$

where

$$\xi = \frac{x}{r},\ D = \frac{E'h}{1-\mu^2},\ K = \frac{h^2}{12r^2},$$

u is the displacements along the x axis; v is the displacements along the tangent line in the direction toward coordinate φ; w is the displacements in the direction

L.Y. Banakh, M.L. Kempner, *Vibrations of Mechanical Systems with Regular Structure*, Foundations of Engineering Mechanics, DOI 10.1007/978-3-642-03126-7_11,

Fig. 11.1 Cylindrical shell

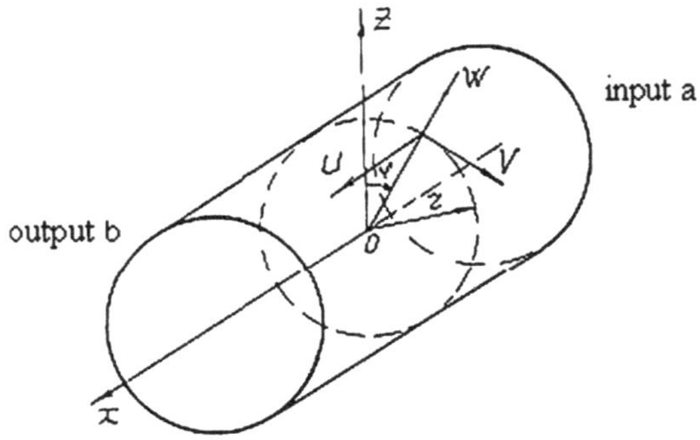

of the external normal to the shell; E', μ, ρ are, respectively, the elasticity module, Poisson's coefficient, and shell material density; r is the curvature radius of the middle surface; h is the shell thickness.

When studying the vibrations of a ribbed shell, it is necessary to have expressions for the generalized forces acting along the coordinate lines at the locations where the shell is connected to the ribbing.

For ring ribbings ($\xi = \text{const}$) (Fig. 11.2a), the forces will be:

$$\begin{aligned} T_1 &= \frac{D}{r}\left[\frac{\partial u}{\partial \varphi} + \mu\left(\frac{\partial v}{\partial \varphi} + w\right)\right] \\ T_{12}^{\circ} &= \frac{Dh(1-\mu)}{2r}\left[\frac{\partial u}{\partial \varphi} + (1+4\kappa)\frac{\partial v}{\partial \xi} - 4\kappa\frac{\partial^2 w}{\partial \xi\,\partial \varphi}\right], \\ M_1 &= 2D\kappa\left[-\frac{\partial^2 w}{\partial \varphi^2} - \mu\left(\frac{\partial^2 w}{\partial \varphi^2} - \frac{\partial v}{\partial \varphi}\right)\right], \\ N^{\circ} &= \frac{D\kappa}{r}\left[-\frac{\partial^3 w}{\partial \xi^3} - (2-\mu)\left(\frac{\partial^3 w}{\partial \xi\,\partial \varphi^2} - \frac{\partial^2 v}{\partial \xi\,\partial \varphi}\right)\right]. \end{aligned} \tag{11.2}$$

For longitudinal ribbings (φ =const) (Fig. 11.2b), the forces are:

$$\begin{aligned} T_2 &= \frac{D}{r}\left(\frac{\partial u}{\partial \varphi} + \omega + \mu\frac{\partial v}{\partial \xi}\right), \\ T_{21} &= \frac{D}{r}\left(\frac{\partial u}{\partial \varphi} + \frac{\partial v}{\partial \xi}\right), \\ M_2 &= Dk\left(\frac{\partial v}{\partial \varphi} - \frac{\partial^2 w}{\partial \varphi^2} - \mu\frac{\partial^2 w}{\partial \xi^2}\right), \end{aligned} \tag{11.3}$$

$$N_2^{\circ} = \frac{Dk}{r}\left[\frac{\partial^2 v}{\partial \varphi^2} + 2(1-\mu)\frac{\partial^2 v}{\partial \xi^2} - \frac{\partial^3 w}{\partial \varphi^3} - (2-\mu)\frac{\partial^3 w}{\partial \xi^2 \partial \varphi}\right].$$

In Eqs. (11.2) and (11.3): T_1 and T_2 are the normal forces, N_1° and N_2° the lateral forces, T_{12}° and T_{21} the shifting forces, and M_1 and M_2 the bending moments.

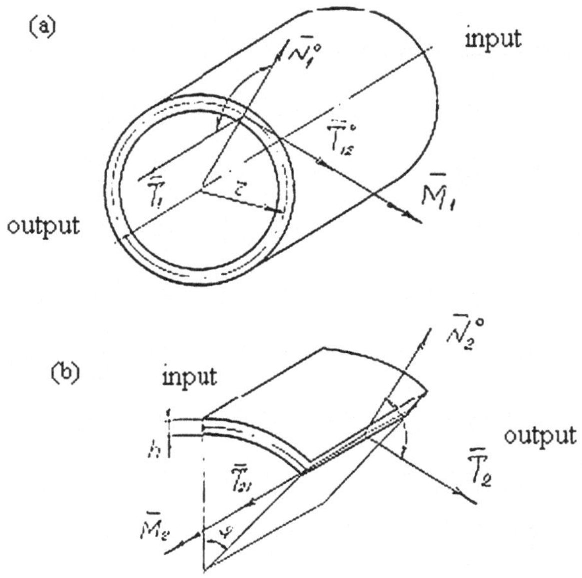

Fig. 11.2 Strains in middle surface

Let us recall that the forces T_{21}°, N_1°, and N_2° are to be understood as a generalization in the sense of Kirchhoff.

Let us now move on to the study of the vibrations of supported shells. Let us first determine the dynamic characteristics of the shell (Sects. 11.2 and 11.3), the ribbings (Sects. 11.4 and 11.5), and then (Sects. 11.6 and 11.7) consider several problems with joint vibrations.

11.2 Dynamic Stiffness and Transition Matrix for Closed Cylindrical Shells

The closed cylindrical shell always has a symmetry plane Π (Fig. 11.3) and is a reflection symmetry element.

We will also treat it as axial-symmetric. The procedure for determining the dynamic stiffness of a cylindrical shell is practically the same as for a rectangular plate supported by a hinge on two opposite sides (see an example of calculation Sect. 6.1). The only difference is that instead of one differential equation in finite differences describing the vibrations of the plate, it is necessary to consider the three equations.

The particular solution of the system of Eqs. (11.1) is sought as a products of the functions of three parameters

$$
\begin{aligned}
u(\xi,\varphi,\omega) &= Ce^{\lambda\xi}\sin s\varphi \sin\omega t,\\
v(\xi,\varphi,\omega) &= Be^{\lambda\xi}\cos s\varphi \sin\omega t,\\
w(\xi,\varphi,\omega) &= Ae^{\lambda\xi}\sin s\varphi \sin\omega t,
\end{aligned}
\tag{11.4}
$$

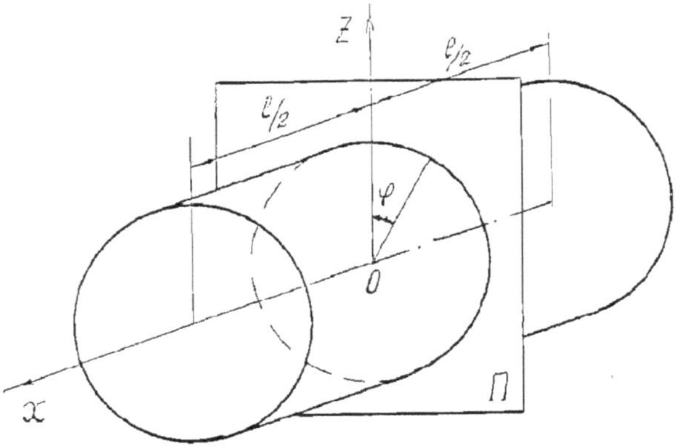

Fig. 11.3 A closed cylindrical shell and its symmetry plane

where s is the number of the nodal lines minus one in the direction of the x axis and ω is the vibration frequency.

Let us substitute (11.4) into (11.1). After that, we obtain a characteristic equation of eighth-order and determine the arbitrary constants A, B, C. Finally, the general solution for system (11.1), which contains eight arbitrary constants $C_1, C_2, \ldots, C_8$·, is found. These constants are expressed by the boundary conditions at the ends of the shell. Then from Eqs. (11.2) the forces acting on a shell input and on an output are defined. Setting the displacements equal to $1 \sin s\varphi$ or $1 \cos s\varphi$, we obtain the dynamic stiffness of the shell we are looking for.

Let us introduce displacement and force vectors:

1. Displacement vector

$$\mathbf{q}(\xi,\varphi) = [u(\xi,\varphi), v(\xi,\varphi), w(\xi,\varphi), \vartheta(\xi,\varphi)]^T, \tag{11.5}$$

where $\vartheta = -\frac{1}{r}\frac{\partial w}{\partial \xi}$ is the inclination angle of the shell section, normal to the x axis, with respect to the coordinate line φ.

Let us place the origin of the coordinates in plane П. Then at $\xi = \xi_1 = \frac{l}{2r}$, we obtain the displacement vector $\mathbf{q}(\xi_1) = \mathbf{q}_b$ at the output from the shell, and at $\xi = -\xi_1$ we will have the displacement vector $\mathbf{q}(-\xi_1) = \mathbf{q}_a$ at the input to the shell.

Let us note the following equalities:

$$\left.\begin{aligned} u(\xi,\varphi) &= u(\xi) \\ w(\xi,\varphi) &= w(\xi) \\ \vartheta(\xi,\varphi) &= \vartheta(\xi) \end{aligned}\right\} \times \sin s\varphi \qquad v(\xi,\varphi) = v(\xi)\cos s\varphi \tag{11.6}$$

2. Force vector

$$\mathbf{Q}(\xi,\varphi) = [T_1(\xi,\varphi), T_{12}^{\circ}(\xi,\varphi), N_1^{\circ}(\xi,\varphi), M_1(\xi,\varphi)]^T \tag{11.7}$$

At $\xi = \xi_1$ we have $\mathbf{Q}(\xi_1, \varphi) = \mathbf{Q}_b$, at $\xi = -\xi_1 \mathbf{Q}(-\xi_1, \varphi) = \mathbf{Q}_a$.
Analogously to Eq. (11.6), we will write the following equations:

$$\left.\begin{aligned} T_1(\xi,\varphi) &= T_1(\xi) \\ N_1^\circ(\xi,\varphi) &= N_1^\circ(\xi) \\ M_1(\xi,\varphi) &= M_1(\xi) \end{aligned}\right\} \times \sin s\varphi$$
$$T_{12}^\circ(\xi,\varphi) = T_{12}^\circ(\xi)\cos s\varphi$$

We will determine the dynamic stiffness of the shell as forces value caused by harmonic excitation with the amplitude equal to one.

$$\begin{aligned} -\mathbf{Q}(-\xi_1) &= \mathbf{R}_{aa}\mathbf{q}(-\xi_1); \\ -\mathbf{Q}(-\xi_1) &= \mathbf{R}_{ab}\mathbf{q}(\xi_1); \\ \mathbf{Q}(\xi_1) &= \mathbf{R}_{ba}\mathbf{q}(-\xi_1); \\ \mathbf{Q}(\xi_1) &= \mathbf{R}_{bb}\mathbf{q}(\xi_1), \end{aligned} \tag{11.8}$$

where $\mathbf{R}_{aa}$, $\mathbf{R}_{ab}$, $\mathbf{R}_{ba}$, $\mathbf{R}_{bb}$ are, respectively, the input, inverse collateral, direct collateral and output dynamic stiffness matrices (all fourth order).

All four matrices can be expressed by the sum or the difference of two matrices (see Eq. (8.1))

$$\begin{aligned} \mathbf{R}_{bb} &= \frac{1}{2}(\bar{\mathbf{R}}_{bb} + \bar{\bar{\mathbf{R}}}_{bb}); \\ \mathbf{R}_{aa} &= \mathbf{I}\mathbf{R}_{bb}\mathbf{I}; \\ \mathbf{R}_{ba} &= \frac{1}{2}(\bar{\mathbf{R}}_{bb} - \bar{\bar{\mathbf{R}}}_{bb})\mathbf{I}; \\ \mathbf{R}_{ab} &= \mathbf{R}T_{ba} = \frac{1}{2}\mathbf{I}(\bar{\mathbf{R}}_{bb} - \bar{\bar{\mathbf{R}}}_{bb}); \end{aligned} \tag{11.9}$$

$$\mathbf{I} = \begin{bmatrix} -1 & 0 & 0 & 0 \\ 0 & 1 & 0 & 0 \\ 0 & 0 & 1 & 0 \\ 0 & 0 & 0 & -1 \end{bmatrix}. \tag{11.10}$$

Using the Eqs. (11.9), we shall define the blocks of a transition matrix for a cylindrical shell:

$$\begin{bmatrix} \mathbf{q}(\xi_1) \\ \mathbf{Q}(\xi_1) \end{bmatrix} = \begin{bmatrix} \mathbf{W}_{11} & \mathbf{W}_{12} \\ \mathbf{W}_{21} & \mathbf{W}_{22} \end{bmatrix} \begin{bmatrix} \mathbf{q}(-\xi_1) \\ \mathbf{Q}(-\xi_1) \end{bmatrix},$$

where blocks $\mathbf{W}_{ij}$ defined from Eqs. (5.16), and (5.17).

Note that the use of the reflection symmetry of the shell has allowed us to reduce the order of the matrix from eight to four.

11.3 Dynamic Stiffness and Transition Matrix for Cylindrical Panel

Let us consider a cylindrical panel (Fig. 11.4) with boundary conditions of the first type (Navies conditions) along two opposite edges *ab* and *cd*. The panel is also a reflection symmetry element with a plane of symmetry Π.

We will search for a particular solution of system (11.1) in a form satisfying the boundary conditions of the first type

$$\begin{aligned} u(\gamma,\xi,\omega) &= Ce^{\lambda\varphi}\cos\alpha\xi\,\sin\omega t,\\ v(\gamma,\xi,\omega) &= Be^{\lambda\varphi}\sin\alpha\xi\,\sin\omega t,\\ w(\gamma,\xi,\omega) &= Ae^{\lambda\varphi}\sin\alpha\xi\,\sin\omega t, \end{aligned} \tag{11.11}$$

where $\xi = \frac{x}{r}$, $\alpha = \frac{m\pi r}{l}$, l is the length of the panel, $m = 0, 1, 2, \ldots$ is the number of node lines in a circular direction.

The displacement vector is

$$\mathbf{q}(\gamma,\xi) = [u(\gamma,\xi), v(\gamma,\xi), \omega(\gamma,\xi), \vartheta(\gamma,\xi)]^T, \tag{11.12}$$

where $\theta = \frac{1}{r}(\frac{\partial w}{\partial\varphi} - v)$ is the angle of rotation of the shell section around the x axis.

The vertex angle of the panel is φ_0 (Fig. 11.4). The coordinate of the input section of the panel is $\varphi = -\frac{\varphi_0}{2}$ and of the output section $\varphi = \frac{\varphi_0}{2}$.

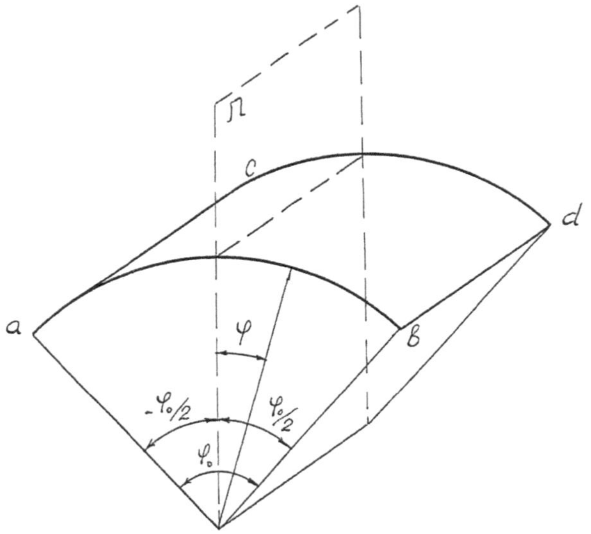

Fig. 11.4 A cylindrical panel and its plane of symmetry

We will write the displacements in (11.11) in the form

$$\begin{aligned} u(\varphi,\xi) &= u(\varphi)\cos\alpha\xi, \\ \left.\begin{aligned} v(\varphi,\xi) &= v(\varphi) \\ \omega(\varphi,\xi) &= \omega(\varphi) \\ \theta(\varphi,\xi) &= \theta(\varphi) \end{aligned}\right\} &\times \sin\alpha\xi. \end{aligned} \tag{11.13}$$

The force vector is

$$\mathbf{Q}(\varphi,\xi) = [T_2(\varphi,\xi), T_{21}^{\circ}(\varphi,\xi), N_2^{\circ}(\varphi,\xi), M_2(\varphi,\xi)]^T, \tag{11.14}$$

$$\begin{aligned} T_2(\varphi,\xi) &= T_2(\varphi)\cos\alpha\xi \\ \left.\begin{aligned} T_{21}(\varphi,\xi) &= T_{21}(\varphi) \\ N_2^{\circ}(\varphi,\xi) &= N_2^{\circ}(\varphi) \\ M_2(\varphi,\xi) &= M_2(\varphi) \end{aligned}\right\} &\times \sin\alpha\xi. \end{aligned} \tag{11.15}$$

The dynamic stiffness of the panel and its transition matrix are determined from Eqs. (11.8)–(11.10). However, matrix **I** now has another view

$$\mathbf{I} = \begin{bmatrix} \mathbf{I} & \mathbf{0} & \mathbf{0} & \mathbf{0} \\ \mathbf{0} & -\mathbf{I} & \mathbf{0} & \mathbf{0} \\ \mathbf{0} & \mathbf{0} & \mathbf{I} & \mathbf{0} \\ \mathbf{0} & \mathbf{0} & \mathbf{0} & -\mathbf{I} \end{bmatrix}. \tag{11.16}$$

All matrices are of the fourth order.

11.4 Dynamic Stiffness and Transition Matrices for Circular Ring with Symmetric Profile

Let us study the spatial vibrations of a ring. It is assumed that a ring can be considered as a curvilinear rod for which the technical theory of bending is valid. It is assumed that a ring has a symmetric profile. One of the main axes of the cross section is perpendicular to the plane of the ring.

Figure 11.5 shows a ring, its main sizes, and the coordinate system. The numbers mark the directions of the displacements in the gravity center for the ring sections and the rotation angle of the section with respect to the coordinate line φ.

The equations for the displacement vector are provided below:

$$\mathbf{q} = \begin{bmatrix} u(\varphi) \\ v(\varphi) \\ w(\varphi) \\ \vartheta(\varphi) \end{bmatrix} = \begin{bmatrix} u\sin s\varphi \\ v\sin s\varphi \\ w\sin s\varphi \\ \vartheta\sin s\varphi \end{bmatrix}, \tag{11.17}$$

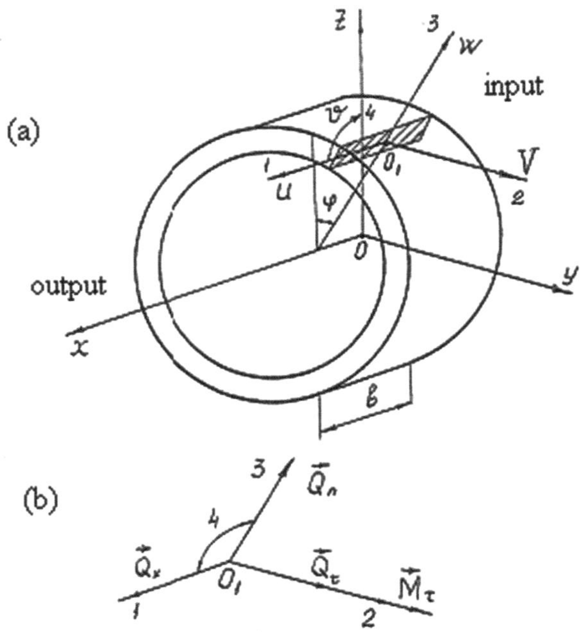

Fig. 11.5a, b A circular cylindrical ring (**a**) and coordinate system (**b**)

The force vector is

$$\mathbf{Q} = \begin{bmatrix} Q_x \sin s\varphi \\ Q_y \sin s\varphi \\ Q_z \sin s\varphi \\ M_y \sin s\varphi \end{bmatrix}. \tag{11.18}$$

The dynamic stiffness matrix for the ring is

$$\mathbf{R}^* = \begin{bmatrix} \mathbf{R}_{11} & 0 & 0 & \mathbf{R}_{14} \\ 0 & \mathbf{R}_{22} & \mathbf{R}_{23} & 0 \\ 0 & \mathbf{R}_{32} & \mathbf{R}_{33} & 0 \\ \mathbf{R}_{41} & 0 & 0 & \mathbf{R}_{44} \end{bmatrix} \tag{11.19}$$

This matrix was obtained from Eq. (7.16) at $\Omega = 0$ and $n = s$.
The transition matrix of the ring is:

$$\begin{bmatrix} \mathbf{q}_b \\ \mathbf{Q}_b \end{bmatrix} = \mathbf{W} \begin{bmatrix} \mathbf{q}_a \\ \mathbf{Q}_a \end{bmatrix}. \tag{11.20}$$

Thereby

$$\mathbf{W} = \begin{bmatrix} \mathbf{E} & \mathbf{0} \\ \mathbf{R}^* & \mathbf{E} \end{bmatrix}. \tag{11.21}$$

Both in the case of a shell and for a circular ring, it is necessary to consider the second type for displacement and forces vectors. For that in Eqs. (11.18), (11.19), it is necessary to replace $\sin s\varphi$ with $\cos s\varphi$ and, conversely, to replace s with –s in (11.20) and (11.21).

11.5 Vibrations of Cylindrical Shell with Ring Ribbing under Arbitrary Boundary Conditions

The cylindrical shell with ring ribbings (Fig. 11.6a) is a regular system with reflection symmetry elements. The distance between the ribbings is the same and is equal to l. The stiffness and inertia parameters of the first and last ribbings are two times less than the parameters of the intermediate ribbings.

As was shown in Chap. 8 for a system with reflection symmetry elements, it is most convenient to use mixed matrices. The application of this method makes it possible to reduce twice the order of the matrices for eigenvalues problem.

Let us consider a regular element of the system shown in Fig. 11.6b. This element consists of two ribbings and a section of the cylindrical shell. The stiffness and inertia parameters of the ribbings coincide with the same parameters of the end ribbings, and therefore there are half as many corresponding parameters of the intermediate ribbings. We will take into account the ribbing's end thickness (b) and the eccentricity (e) with which the ribbing is fixed to the shell [59]; the eccentricity can be positive or negative.

We will determine the dynamic stiffness matrix of the regular element as a sum of the dynamic stiffness matrices of the ring ribbing and the section of the shell. To this end, we must find first this matrix for the shell in the axes passing through the gravity center of the ribbing cross sections.

The dynamic stiffness matrix of the shell is determined from Eq. (11.9).

Let us now transition to coordinate systems passing through the gravity center of the left and right ribbings (Fig. 11.7).

In the new axes, the dynamic stiffness matrix of the shell is determined from Equality (A.5)

$$\mathbf{R}^{\circ}_{aa} = \mathbf{\Gamma}^T_a \mathbf{R}_{aa} \mathbf{\Gamma}_a; \mathbf{R}^{\circ}_{ab} = \mathbf{\Gamma}^T_b \mathbf{R}_{ab} \mathbf{\Gamma}_a;$$
$$\mathbf{R}^{\circ}_{ba} = \mathbf{\Gamma}^T_a \mathbf{R}_{ba} \mathbf{\Gamma}_b; \mathbf{R}^{\circ}_{bb} = \mathbf{\Gamma}^T_b \mathbf{R}_{bb} \mathbf{\Gamma}_b,$$

N are the transformation matrices determined from Eqs. (A.6) and (2.27).

From Eq. (A.1) we have

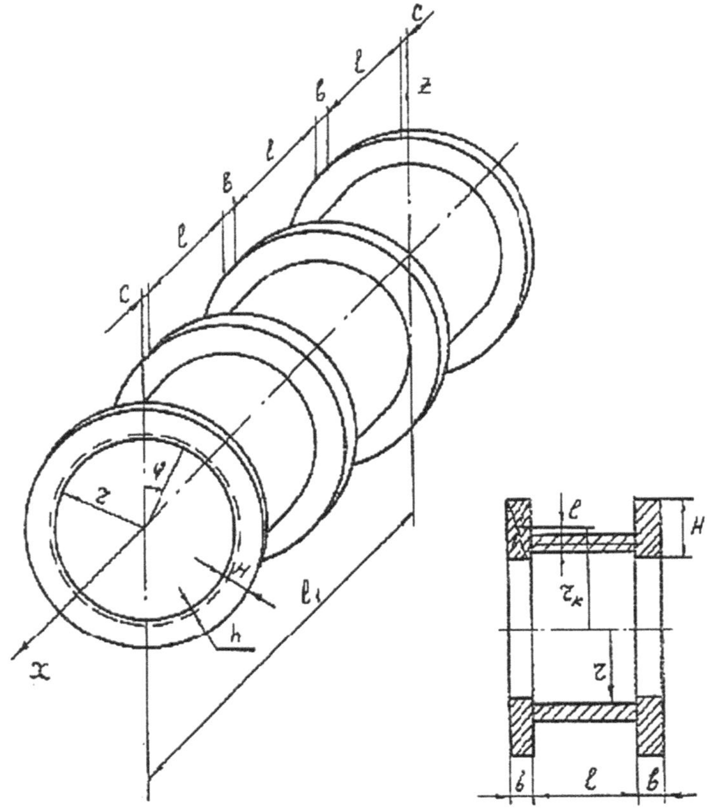

Fig. 11.6 A cylindrical shell with circular ribbings

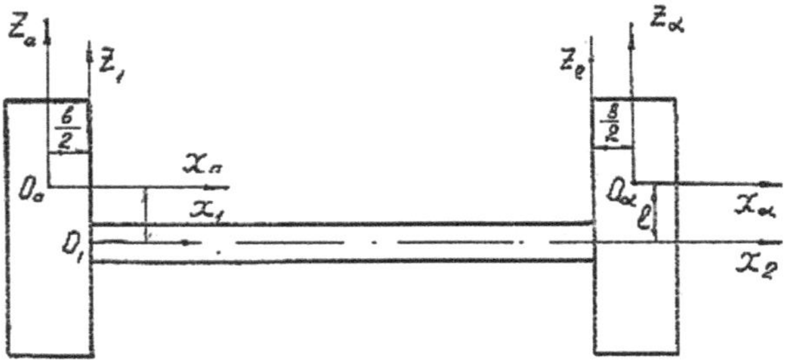

Fig. 11.7 Relative position of ring ribbings and longitudinal section of a shell

$$\mathbf{q}_1 = \mathbf{L}_a^T \mathbf{q}_a.$$
$$\mathbf{q}_2 = \mathbf{L}_b^T \mathbf{q}_b,$$

where $\mathbf{N}_a = \mathbf{L}_a^{-1}$, $\mathbf{N}_b = \mathbf{L}_b^{-1}$, and $\mathbf{q}_{a,}\mathbf{q}_b$ are six-dimensional displacements vectors in the axes $O_aX_aY_aZ_a$ and $O_bX_bY_bZ_b$ and $\mathbf{q}_{1,}\mathbf{q}_2$ are the vectors at the input and the output.

Let us now determine matrix $\mathbf{N}_a$. From Eq. (A.1):

$$\mathbf{N}_a = \begin{bmatrix} \boldsymbol{\theta} & \boldsymbol{\theta}\mathbf{V}_1 \\ \mathbf{0} & \boldsymbol{\theta} \end{bmatrix}. \qquad \text{(A.2), (2.27)}$$

In our case $\boldsymbol{\theta} = \mathbf{E}$, therefore

$$\mathbf{N}_a^T = \begin{bmatrix} \mathbf{E} & \mathbf{0} \\ \mathbf{V}^T & \mathbf{E} \end{bmatrix}, \tag{11.22}$$

$$\mathbf{V} = \begin{bmatrix} 0 & e & 0 \\ -e & 0 & -\frac{b}{2} \\ 0 & \frac{b}{2} & 0 \end{bmatrix}. \tag{11.23}$$

However, not all six component vectors $\mathbf{q}_a$ are independent. For a circular ring, the rotation angles θ_x and θ_z can be expressed through displacements u, v, and w [112]

$$\begin{aligned} \theta_x &= -\frac{v}{rk} + \frac{1}{rk}\frac{\partial w}{\partial \varphi}, \\ \theta_z &= -\frac{1}{rk}\frac{\partial u}{\partial \varphi}. \end{aligned} \tag{11.24}$$

Setting

$$\begin{aligned} u(\varphi) &= u \sin s\varphi; \quad & Q_x(\varphi) &= Q_x \sin s\varphi \\ v(\varphi) &= v \sin s\varphi; \quad & \vartheta(\varphi) &= \vartheta \sin s\varphi \quad , \\ w(\varphi) &= w \sin s\varphi; \quad & \theta_z(\varphi) &= \theta_z \sin s\varphi \end{aligned} \tag{11.25}$$

we can reduce the order of matrix $\mathbf{N}_a$ to four using (11.24):

$$\mathbf{N}_a = \begin{bmatrix} 1 & -\frac{sb}{2rk} & 0 & 0 \\ 0 & 1-\frac{e}{rk} & 0 & 0 \\ 0 & \frac{se}{rk} & 1 & 0 \\ -e & 0 & -\frac{b}{2} & 1 \end{bmatrix} \tag{11.26}$$

And analogously

$$\mathbf{N}_b = \begin{bmatrix} 1 & \frac{sb}{2rk} & 0 & 0 \\ 0 & 1-\frac{e}{rk} & 0 & 0 \\ 0 & \frac{se}{rk} & 1 & 0 \\ -e & 0 & \frac{b}{2} & 1 \end{bmatrix} \tag{11.27}$$

It is obvious that

$$\mathbf{N}_a = \mathbf{I}\mathbf{N}_b\mathbf{I},$$
$$\mathbf{N}_a^T = \mathbf{I}\mathbf{N}_b^T\mathbf{I}.$$

The dynamic stiffness matrix of the shell in axis $O_aX_aY_aZ_a$ and $O_bX_bY_bZ_b$ are:

$$\begin{aligned}
\mathbf{R}_{bb}^{\circ} &= \frac{1}{2}\mathbf{N}^T{}_b(\bar{\mathbf{R}} + \bar{\bar{\mathbf{R}}})\mathbf{N}_b \\
\mathbf{R}_{aa}^{\circ} &= \mathbf{I}\mathbf{R}_{bb}^{\mathbf{0}}\mathbf{I}, \\
\mathbf{R}_{ba}^{\circ} &= \frac{1}{2}\mathbf{N}_b(\bar{\mathbf{R}} - \bar{\mathbf{R}})\mathbf{I}\mathbf{N}_a^T, \\
\mathbf{R}_{ab}^{\circ} &= [\mathbf{R}_{ba}^{\circ}]^T.
\end{aligned} \tag{11.28}$$

The dynamic stiffness of the regular element is

$$\begin{aligned}
\mathbf{R}_{aa}^{el} &= \mathbf{R}^r + \mathbf{R}_{aa}^{\circ} \\
\mathbf{R}_{ab}^{el} &= \mathbf{R}_{b\alpha}^{\circ}, \\
\mathbf{R}_{ba}^{el} &= \mathbf{R}_{ba}^{\circ} \\
\mathbf{R}_{bb}^{el} &= \mathbf{R}^r + \mathbf{R}_{bb}^{\circ},
\end{aligned} \tag{11.29}$$

where $\mathbf{R}^r$ is the dynamic stiffness matrix of the ring.

Let us now rearrange the components of the vectors in such a way that the first are components of the displacements of the first group (v and w) and the second are components of the second group (u, ϑ).

$$\mathbf{q}_1 = [v, w]^T, \mathbf{q}_2 = [u, \vartheta]^T,$$

and

$$\mathbf{Q}_1 = [\mathbf{T}_{12}^{\circ}, \mathbf{N}_1^0]^T, \mathbf{Q}_2 = [\mathbf{T}_1, \mathbf{M}_1]^T.$$

By using the elements of matrix (11.29), let us write in block form the dependence between the forces and the displcements.

$$\begin{bmatrix} -\mathbf{Q}_{1a} \\ -\mathbf{Q}_{2a} \\ \mathbf{Q}_{1b} \\ \mathbf{Q}_{2b} \end{bmatrix} = \begin{bmatrix} \mathbf{R}_{11}^{el} & -\mathbf{R}_{12}^{el} & \mathbf{B}_{11}^{el} & \mathbf{B}_{12}^{el} \\ \mathbf{R}_{21}^{el} & \mathbf{R}_{22}^{el} & \mathbf{B}_{21}^{el} & \mathbf{B}_{22}^{el} \\ \mathbf{B}_{11}^{el} & -\mathbf{B}_{12}^{el} & \mathbf{R}_{11}^{el} & \mathbf{R}_{12}^{el} \\ \mathbf{B}_{21}^{el} & \mathbf{B}_{22}^{el} & \mathbf{R}_{21}^{el} & \mathbf{R}_{22}^{el} \end{bmatrix} \begin{bmatrix} \mathbf{q}_{1a} \\ \mathbf{q}_{2a} \\ \mathbf{q}_{1b} \\ \mathbf{q}_{2b} \end{bmatrix}$$

The characteristic equation of the mixed matrix is

$$\text{Det}\,[\mathbf{N} - \lambda\mathbf{D}^*] = 0$$

and the elements of matrices **N** and **P*** are determined from Eq. (7.3). Both these two matrices are symmetric and of the fourth order.

The division of a ribbed shell into regular elements permits one to facilitate considerably a solution to the problem of finding the natural frequencies. This is particularly important when some additional elastic-mass inclusions (masses, rods, springs with masses, blades, etc.) have been added to the ring ribbing. In this case, the dynamic stiffness of the ring are determined by taking into account these inclusions. Obviously, the condition for the identity of all elements remains.

In the often case of boundary conditions of the first type, the characteristic values of the ribbed shell is determined from Eq. (7.43)

$$\lambda = \cos \frac{\pi m}{n} (m = 1, 2, \ldots, n - 1).$$

11.6 Vibrations of Cylindrical Shell with Longitudinal Ribbing of Nonsymmetric Profile

The free vibrations of a closed cylindrical shell supported by longitudinal ribbings of stiffness (Fig. 11.8a,b) can be treated analogously. The structure itself is a system with rotary symmetry; however, the ribbing can have an asymmetric profile.

A most complete solution of this problem is provided in [73] and [59]. However, only ribbings with a symmetric profile were considered in these studies. Thus the axis of symmetry coincided with the radius of the shell. Here we consider a more general case.

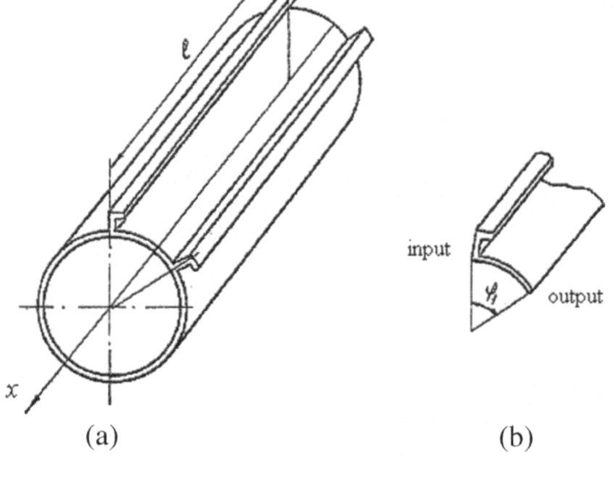

Fig. 11.8a, b A shell with longitudinal ribbings of stiffness (**a**) and a regular element (**b**)

The shell with the ribbings is presented in Fig. 11.8a. Figure 11.8b shows a regular element of the system consisting of one ribbing and a cylindrical panel. Both external and internal ribs of the shell is possible.

11.6.1 Dynamic Stiffness and Transition Matrices for Longitudinal Stiffening Ribs

A longitudinal ribbing is a rectilinear beam with a hinged support at its ends (Fig. 11.9a). It may have no symmetrical profile. In studying ribbing vibrations, the inertia of the rotation and the displacement deformation are taken into account.

Let us introduce the displacement vector

$$\mathbf{q} = \begin{bmatrix} u \sin \alpha\xi \\ v \sin \alpha\xi \\ \omega \sin \alpha\xi \\ \theta \sin \alpha\xi \end{bmatrix},$$

where u, v, and w are the displacement of the gravity center of the ribbing in the coordinate system OXY_1Z_1 (Fig. 11.9b), Y_1, Z_1 are the principal axes of the cross section of the ribbing, θ is the rotation angle of the cross section of the ribbing about the x axis, and ξ and α are determined from Eq. (11.11).

Force vector $\mathbf{Q}$ corresponds to displacement vector $\mathbf{q}$.

$$\mathbf{Q} = \begin{bmatrix} Q_x \sin \alpha\xi \\ Q_{y1} \sin \alpha\xi \\ Q_{z1} \sin \alpha\xi \\ M_x \cos \alpha\xi \end{bmatrix},$$

The matrix of dynamic stiffness of the ribbing is

$$\mathbf{R}_{rib} = \begin{bmatrix} R_{11} & 0 & 0 & 0 \\ 0 & R_{22} & 0 & 0 \\ 0 & 0 & R_{33} & 0 \\ 0 & 0 & 0 & R_{44} \end{bmatrix},$$

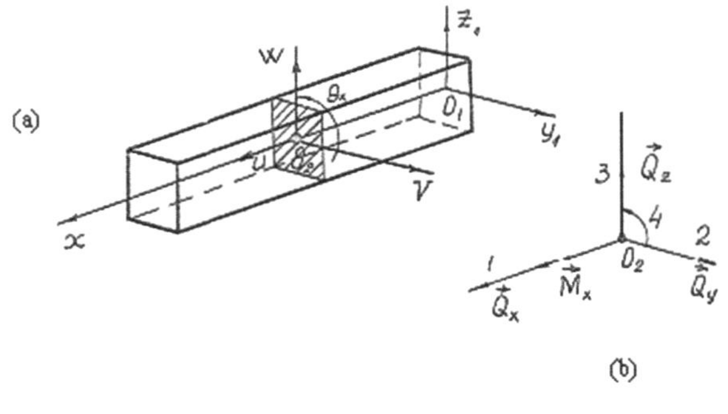

Fig. 11.9a, b Coordinate axes of a longitudinal ribbing

where

$$R_{11} = -E'F\left[\left(\frac{\alpha}{r}\right)^2 - \frac{\rho_1\omega^2}{E'}\right],$$

$$R_{22} = E'J_z\left[\left(\frac{\alpha}{r}\right)^4 - \frac{S_1}{E'}\left(\frac{\alpha\omega}{r}\right)^2\left(1+\frac{E'K_{1z}}{G}\right) - \frac{S_1F}{E'J_z}\omega^2\right],$$

$$R_{33} = E'J_y\left[\left(\frac{\alpha}{r}\right)^4 - \frac{S_1}{E'}\left(\frac{\alpha\omega}{r}\right)^2\left(1+\frac{E'K_{1y}}{G}\right) - \frac{S_1F}{E'J_y}\omega^2\right],$$

$$R_{44} = GJ_p\left[\left(\frac{\alpha}{r}\right)^2 - \frac{\rho_1\omega^2}{G}\right],$$

The effect of the displacement has been taken into account in the coefficients K_{1y} and K_{1z}.

The transition matrix of the ribbing is

$$\mathbf{W}_{rib} = \begin{bmatrix} \mathbf{E} & 0 \\ \mathbf{R}_{rib} & \mathbf{E} \end{bmatrix}.$$

Its order is (8×8).

11.6.2 Numerical Calculation of Shell with Longitudinal Ribbing

Unlike the problem considered in the preceding subsection, the regular element in this problem is not reflection symmetric [59].

Here we use the results obtained in Chap. 7 (Sect. 7.1).

The transition matrix of the regular element shown in Fig. 11.8b can be presented in the form of the multiplication of two matrices

$$\mathbf{W} = \mathbf{W_0}\mathbf{W}'_{\text{rib}}. \tag{11.30}$$

Here $\mathbf{W}_0$ is the transition matrix for a cylindrical panel, matrix **W0** was determined in Sect. 11.3. $\mathbf{W}'_{\text{rib}}$ is the transition matrix for the ribbing; it is necessary to take into account the rotation α between the ribbing and the panel. Then from Eqs. (2.27) and (9.37):

$$\mathbf{W}'_{\text{rib}} = \mathbf{N}^T\mathbf{W}_{\text{rib}}\mathbf{N}.$$

The natural frequencies are determined from the equation

$$\det[(\mathbf{W} - \mathbf{E})^2 + 4\sin^2\frac{\pi s}{n}\mathbf{W}] = 0$$

($s = 1, 2, \ldots, n/2$; $n =$ the number of elements).

Table 11.1 Frequencies of vibrations of a shell with longitudinal ribbings $\omega_{ms} cek^{-1}$

s \ m	1	2
1	9540	26369
2	4926	14730

Two equal vibration frequencies ω correspond to each pair of integer values m and s, but the natural modes determined from Eq. (7.13) will be different for each of these frequencies.

An example for calculating a shell with outside supports (Fig. 11.10) is provided in Table 11.1.

Dimensions of ribbing and shell (in centimeters): shell radius $r = 4.9$ cm, shell length $l = 39.4$ cm, $h = 0.165$ cm, $H = 0.8$ cm, $e = 0.63$ cm, $b = 0.12$ cm, $b_1 = -0.17$ cm, $d = 0.6$ cm, $h_1 = 0.15$ cm, number of ribbings $n = 15$, density of shell and ribbing material.$\rho = 2.6 \times 10^{-2}$n/cm^3, elasticity module $E = 0.7 \cdot 10^5$n/cm^2.

All frequencies are multiple. In the case of ribbings with a symmetric profile, the phase displacement between the maximum values of the different component of the displacement vectors is equal to zero or is a multiple of $\pi/2$. The presence of a ribbing in an asymmetric profile leads to a phase displacement that differs from the provided values. For example, in the problem under consideration, the phase displacement between the maximal discrete values $\theta[k]$ and $v[k]$ is $\approx 24°$, which corresponds to the angle between two neighboring ribbings ($v[k]$ is measured in the direction of the line tangent to the shell, and k is the number of the ribbing).

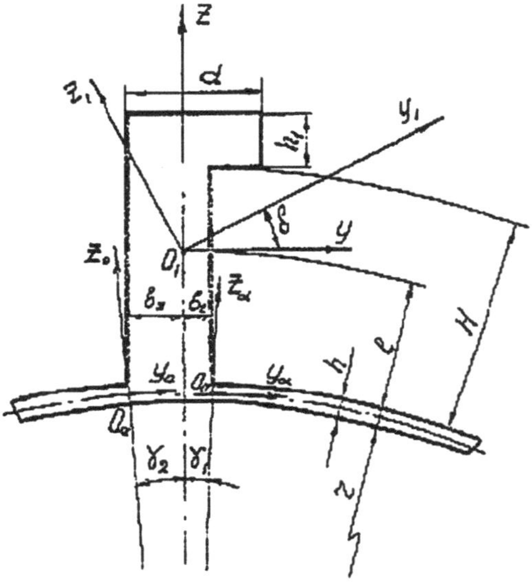

Fig. 11.10 Relative positioning of ribbing section and shell

Appendix A
Stiffness and Inertia Matrices for a Ramified System Consisting of Rigid Bodies Connected by Beam Elements

Let us consider a system consisting of rigid bodies connected by beam elements. In order to obtain the stiffness and inertia matrices for this system, let us consider first a beam element connecting two rigid bodies [15].

Let us choose a stationary coordinate system $(x_1y_1z_1)$ fixed to body 1 (Fig. A.1); the zero of the coordinate system O_1 is located at the center of mass of the body. Let us assume that this body is linked to an elastic beam element; its fixing point O'_1 is at a distance from the center of mass equal to radius vector $\boldsymbol{\rho}_1$. Let us now choose a local coordinate system linked by a beam element $(x'_1y'_1z'_1)$ whose beginning is at the fixing point O'_1 and the coordinate axes run along its main central axes as shown in Fig. A.1. Each node of the system has six degrees of freedom. The ratios between the shifts along the coordinate axes $\mathbf{X}'(x'_1y'_1z'_1)$ and the rotations around them $\boldsymbol{\Psi}$, and the corresponding forces $\mathbf{F}'_{\mathbf{X}}$ and the moments $\mathbf{F}'_{\boldsymbol{\Psi}}$, are written with the help of the stiffness matrix $\mathbf{K}_b$ for the beam element in the local coordinate system

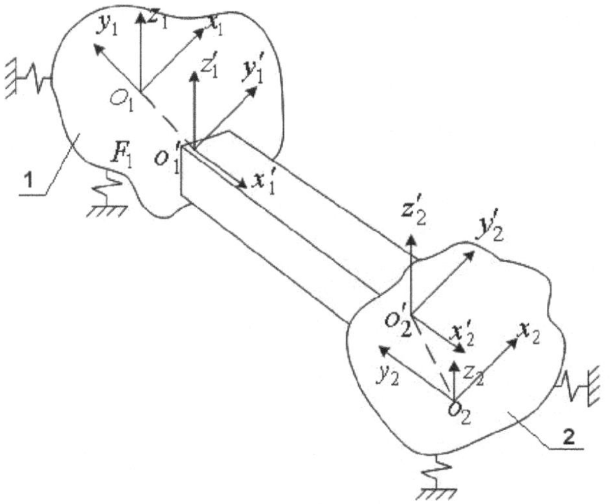

Fig. A.1 Three-dimensional finite beam element and its local coordinates systems

L.Y. Banakh, M.L. Kempner, *Vibrations of Mechanical Systems with Regular Structure*, Foundations of Engineering Mechanics, DOI 10.1007/978-3-642-03126-7,

$$\begin{bmatrix} \mathbf{F}'_{\mathbf{x}} \\ \mathbf{F}'_{\Psi} \end{bmatrix} = \mathbf{K}_b \begin{bmatrix} \mathbf{X}' \\ \Psi \end{bmatrix}, \quad \mathbf{K}_b = \begin{bmatrix} \mathbf{K}_{11b} & \mathbf{K}_{12b} \\ \mathbf{K}_{21b} & \mathbf{K}_{22b} \end{bmatrix}. \tag{A.1}$$

where $\boldsymbol{\Psi} = [\theta_x, \theta_y, \theta_z]$ is the vector of the angular rotations. (Here the first end of the beam reaches body 1 and the second one reaches body 2.)

The relative displacement of the beam element and the body can be determined using the following matrices: $\boldsymbol{\theta}$ is the matrix of the directional cosine angles between axes $(x'y'z')$ and (xyz) for the first and second ends of the beam, respectively

$$\boldsymbol{\theta}^{(i)} = \begin{bmatrix} \alpha_{xx'} & \alpha_{yx'} & \alpha_{zx'} \\ \alpha_{xy'} & \alpha_{yy'} & \alpha_{zy'} \\ \alpha_{xz'} & \alpha_{yz'} & \alpha_{zz'} \end{bmatrix}^{(i)}, \quad (i = 1, 2),$$

and $\boldsymbol{\rho}_1$, $\boldsymbol{\rho}_2$ is the matrix of the shifts of the coordinate system origin for the first and second ends of the beam

$$\boldsymbol{\rho}_1 = \begin{bmatrix} 0 & \rho_z & -\rho_y \\ -\rho_z & 0 & \rho_x \\ \rho_y & -\rho_x & 0 \end{bmatrix}_1,$$

$\rho_x = x - x'$, $\rho_y = y - y'$, $\rho_z = z - z'$ (index "$_1$" has been omitted for the sake of simplicity). Analogous ratios can also be written for matrix $\boldsymbol{\rho}_2$ of the second end of the beam.

Matrix $\boldsymbol{\theta}$ is characterized by the rotation of the coordinate axes of the beam element in relation to the body. It can be represented as a product of the three orthogonal matrices $\boldsymbol{\theta}_1$, $\boldsymbol{\theta}_2$, $\boldsymbol{\theta}_3$ describing the consecutive angles $\theta_x, \theta_y, \theta_z$ of rotation around the coordinate axes

$$\boldsymbol{\theta}_1 = \begin{bmatrix} 1 & 0 & 0 \\ 0 & \cos\theta_x & -\sin\theta_x \\ 0 & \sin\theta_x & \cos\theta_x \end{bmatrix}, \quad \boldsymbol{\theta}_2 = \begin{bmatrix} \cos\theta_y & 0 & \sin\theta_y \\ 0 & 1 & 0 \\ -\sin\theta_y & 0 & \cos\theta_y \end{bmatrix},$$

$$\boldsymbol{\theta}_3 = \begin{bmatrix} \cos\theta_z & -\sin\theta_z & 0 \\ \sin\theta_z & \cos\theta_z & 0 \\ 0 & 0 & 1 \end{bmatrix},$$

$$\boldsymbol{\theta} = \boldsymbol{\theta}_1 \boldsymbol{\theta}_2 \boldsymbol{\theta}_3.$$

If there is no rotation, then $\boldsymbol{\theta} = \mathbf{E}$, where $\mathbf{E}$ is the identity matrix.

In the general case, the shift of the beam in the local coordinate system $(x'y'z')$ can be carried over into a stationary coordinate system $(Oxyz)$ linked to the body using the transformation matrix $\boldsymbol{\Gamma}$, which takes into account both the transfer of the origin of the coordinates and the rotation of the axes. Or in block form

$$\begin{bmatrix} \mathbf{X}' \\ \boldsymbol{\Psi}' \end{bmatrix} = \begin{bmatrix} \boldsymbol{\theta} & \boldsymbol{\theta}\boldsymbol{\rho}_1 \\ \mathbf{0} & \boldsymbol{\theta} \end{bmatrix} \begin{bmatrix} \mathbf{X} \\ \boldsymbol{\Psi} \end{bmatrix} = \boldsymbol{\Gamma}_1 \begin{bmatrix} \mathbf{X} \\ \boldsymbol{\Psi} \end{bmatrix}, \qquad \boldsymbol{\Gamma}_1 = \begin{bmatrix} \boldsymbol{\theta} & \boldsymbol{\theta}_{\rho_1} \\ \mathbf{0} & \boldsymbol{\theta} \end{bmatrix}. \tag{A.2}$$

for the first end of the beam and, analogously, for the second end. Then the transformation matrix for the whole beam has the following block-diagonal form:

$$\boldsymbol{\Gamma} = \begin{bmatrix} \boldsymbol{\Gamma}_1 & \\ & \boldsymbol{\Gamma}_2 \end{bmatrix}.$$

The transformation of the force factors in the node as written in the stationary coordinate system is

$$\begin{bmatrix} \mathbf{F}'_X \\ \mathbf{F}'_\theta \end{bmatrix} = \begin{bmatrix} \boldsymbol{\theta} & \mathbf{0} \\ \boldsymbol{\theta}_\rho & \mathbf{0} \end{bmatrix} \begin{bmatrix} \mathbf{Q}_X \\ \mathbf{Q}_\theta \end{bmatrix}.$$

Analogously for the first and second ends of the beam.

Taking into account that

$$\boldsymbol{\theta}^T\boldsymbol{\theta}^{-T}, \ \boldsymbol{\rho}^T = -\boldsymbol{\rho}, (\boldsymbol{\rho}\,\boldsymbol{\theta})^T = \boldsymbol{\theta}^T\,\boldsymbol{\rho}^T = -\boldsymbol{\theta}^T\,\boldsymbol{\rho},$$

the reverse transformation has the form

$$\begin{bmatrix} \mathbf{Q}_X \\ \mathbf{Q}_\theta \end{bmatrix} = \begin{bmatrix} \boldsymbol{\theta}^T & \mathbf{0} \\ (\boldsymbol{\theta}\boldsymbol{\rho})^T & \boldsymbol{\theta}^T \end{bmatrix} \begin{bmatrix} \mathbf{F}'_X \\ \mathbf{F}'_\theta \end{bmatrix} = \boldsymbol{\Gamma} \begin{bmatrix} \mathbf{F}'_X \\ \mathbf{F}'_\theta \end{bmatrix}. \tag{A.3}$$

In particular, force $\mathbf{F}_X$, which acts on the beam element, generates a moment (at $\boldsymbol{\theta} = \mathbf{E}$) in the (x,y,z) axes:

$$\mathbf{Q}_\theta = \boldsymbol{\rho} \times \mathbf{F}_x.$$

Or in matrix form

$$\mathbf{Q}_\theta = \boldsymbol{\rho}^T\mathbf{F}_x.$$

From (A.1) and (A.3) we find

$$\begin{bmatrix} \mathbf{Q}_X \\ \mathbf{Q}_\theta \end{bmatrix} = \boldsymbol{\Gamma}^T\mathbf{K}_b \begin{bmatrix} \mathbf{X}' \\ \boldsymbol{\Psi}' \end{bmatrix} = \boldsymbol{\Gamma}^T\mathbf{K}_b\boldsymbol{\Gamma} \begin{bmatrix} \mathbf{X} \\ \boldsymbol{\Psi} \end{bmatrix}. \tag{A.4}$$

From which it follows that the stiffness matrix for the beam element has been carried over to node O:

$$\mathbf{K} = \boldsymbol{\Gamma}^T\,\mathbf{K}_\mathrm{b}\,\boldsymbol{\Gamma}. \tag{A.5}$$

Obviously, this transformation can be written in the following way as a block product:

$$\mathbf{K}_{11} = \boldsymbol{\Gamma}_1^T \mathbf{K}_{11b} \boldsymbol{\Gamma}_1\,, \quad \mathbf{K}_{12} = \boldsymbol{\Gamma}_1^T \mathbf{K}_{12b} \boldsymbol{\Gamma}_2\,,$$
$$\mathbf{K}_{22} = \boldsymbol{\Gamma}_2^T \mathbf{K}_{22b} \boldsymbol{\Gamma}_2\,, \quad \mathbf{K}_{12} = \mathbf{K}_{21}^T.$$

Using (A.4), a simple algorithm in analytical form for determining the end-element stiffness matrix can be proposed (Appendix B).

The compliance matrix $\mathbf{e} = \mathbf{K}^{-1}$ can be written in the following form in the stationary coordinate system by taking into account (A.5)

$$\mathbf{e} = \mathbf{K}^{-1} = \boldsymbol{\Gamma}^{-1}\mathbf{e}_{\mathbf{b}}\left(\boldsymbol{\Gamma}^T\right)^{-1}, \quad \boldsymbol{\Gamma}^{-1} = \begin{bmatrix} \boldsymbol{\theta}^T & \\ \boldsymbol{\rho}^T\boldsymbol{\theta}^T & \boldsymbol{\theta}^T \end{bmatrix}. \tag{A.6}$$

Knowing the stiffness (compliance) matrix for each end element, the corresponding matrices for the whole system can be found from the compatibility condition for the forces (shifts) in the decomposition nodes. So the stiffness (compliance) matrix for each *i*th node is obtained as a sum of the matrices for all beams participating in the given node

$$\hat{\mathbf{K}}_{ii} = \sum_{\mathbf{s}} \boldsymbol{\Gamma}_{\mathbf{s}}^T \mathbf{K}_{\mathbf{s}} \boldsymbol{\Gamma}_{\mathbf{s}}, \tag{A.7}$$

where s is the number of end elements included in the *i*th node.

Blocks $\hat{\mathbf{K}}_{ij}$ reflect the links between nodes *i* and *j*

$$\hat{\mathbf{K}}_{ij} = \boldsymbol{\Gamma}_i^T \mathbf{K}_{ij} \boldsymbol{\Gamma}_j.$$

In this way, we found all the blocks that form the stiffness matrix of the end-element model.

Definition A subsystem is called partial if it was obtained by rigid fixing of all remaining subsystems (nodes).

Therefore, we can consider the *i*-th node a partial subsystem that is described by matrix $\mathbf{K}_{ii}$.

Example. Let us consider as an example two flat beams linked at point *C* under a certain angle (e.g., $3\pi/4$) (Fig. A.2)

The general (stationary) coordinate system is chosen at the linkage point as shown in Fig. A.2. Therefore, it coincides with the main inertia axes for beam *A*. For beam *A*, the rotation matrix $\boldsymbol{\theta} = \mathbf{E}$ is at end 2. If, in addition to that, the center of inertia of the cross section coincides with point C, then the displacement matrix $\boldsymbol{\rho}_{\mathbf{2}}^{\mathbf{A}} = \mathbf{E}$ and the transformation matrix $\boldsymbol{\Gamma}_{\mathbf{2}}^{\mathbf{A}} = \mathbf{E}$. End 1 of beam A remains unchanged, so that for it the transformation matrix $\boldsymbol{\Gamma}_{\mathbf{1}}^{\mathbf{A}} = \mathbf{E}$. As regards beam B, its main central axes at end 1 must be turned around the *z* axis at an angle of $3\pi/4$ until they coincide with the general coordinate system. Therefore, the rotation matrix $\boldsymbol{\theta}$ will have an angle $\theta_z = \varphi = 3\pi/4$.

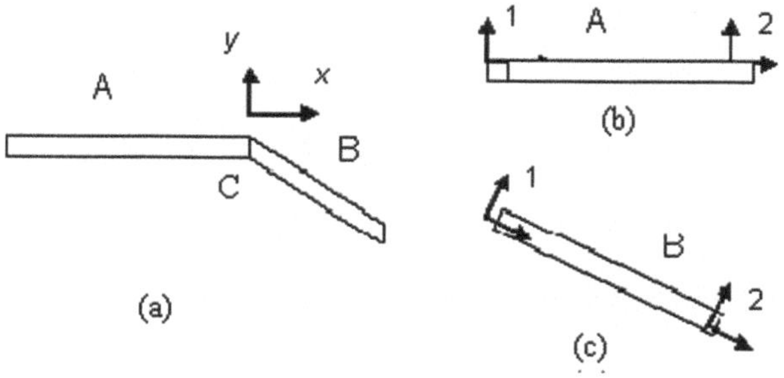

Fig. A.2 Flat beams linked under a certain angle

In order to obtain the stiffness matrix of the linked system, it is necessary to perform certain matrix transformations:

– for beam A, its source stiffness matrix (2.29) is:

$$\mathbf{K}^A = \begin{bmatrix} \mathbf{K}_{11}^A & \mathbf{K}_{12}^A \\ \mathbf{K}_{21}^A & \mathbf{K}_{22}^A \end{bmatrix}.$$

For the first end, its transformation matrix is $\boldsymbol{\Gamma}_1^A = \mathbf{E}$, and for the second end – as we have just shown – also $\boldsymbol{\Gamma}_2^A = \mathbf{E}$. Therefore, the stiffness matrix for beam A will not change since $\boldsymbol{\Gamma}^A = \mathbf{E}$.

– For beam *B*, its source stiffness matrix is analogous to

$$\mathbf{K}^B = \begin{bmatrix} \mathbf{K}_{11}^B & \mathbf{K}_{12}^B \\ \mathbf{K}_{21}^B & \mathbf{K}_{22}^B \end{bmatrix}.$$

For the first end of the rotation matrix

$$\boldsymbol{\theta}_1^B = \begin{bmatrix} \cos\varphi & -\sin\varphi \\ \sin\varphi & \cos\varphi \end{bmatrix}.$$

As we have presumed above, the shift matrix for it is $\boldsymbol{\rho}_1^B = \mathbf{E}$. Then the transformation matrix for the first end is $\boldsymbol{\Gamma}_1^B = \boldsymbol{\theta}_1^B$. End 2, as above, remains unchanged: for this end, the transformation matrix is $\boldsymbol{\Gamma}_2^B = \mathbf{E}$. Therefore, matrix $\mathbf{K}^B$ must be multiplied from the left and the right into the following transformation matrix

$$\boldsymbol{\Gamma}^B = \begin{bmatrix} \boldsymbol{\theta}_1^B & \\ & \mathbf{E} \end{bmatrix},$$

and into its transposed matrix

$$\hat{\mathbf{K}}^{\mathbf{B}} = \begin{bmatrix} \mathbf{A}_1^{B\mathbf{tr}} & \\ & \mathbf{E} \end{bmatrix} \begin{bmatrix} \mathbf{K}_{11}^B & \mathbf{K}_{12}^B \\ \mathbf{K}_{21}^B & \mathbf{K}_{22}^B \end{bmatrix} \begin{bmatrix} \mathbf{A}_1^B & \\ & \mathbf{E} \end{bmatrix} =$$
$$= \begin{bmatrix} \mathbf{A}_1^{B\mathbf{tr}} \mathbf{K}_{11}^B \mathbf{A}_1^B & \mathbf{A}_1^{B\mathbf{tr}} \mathbf{K}_{12}^B \\ \mathbf{A}_1^B \mathbf{K}_{21}^B & \mathbf{K}_{22}^B \end{bmatrix}.$$

And finally, the stiffness matrix for the linked system in Fig. A.2a will take the following form:

$$\mathbf{K} = \begin{bmatrix} \mathbf{K}_{11}^A & \mathbf{K}_{12}^A & \\ \mathbf{K}_{21}^A & \mathbf{K}_{22}^A + \boldsymbol{\theta}_1^{Btr} \mathbf{K}_{11}^B \boldsymbol{\theta}_1^B & \boldsymbol{\theta}_1^{Btr} \mathbf{K}_{12}^B \\ & \mathbf{K}_{21}^B \boldsymbol{\theta}_1^B & \mathbf{K}_{22}^B \end{bmatrix}. \tag{2.27}$$

Appendix B
Stiffness Matrix for Spatial Finite Element

General view for stiffness matrix for spatial beam finite element is (Fig. A.1):

$$\mathbf{K} = \begin{bmatrix} \mathbf{K}_{11} & \mathbf{K}_{12} \\ \mathbf{K}_{12}^T & \mathbf{K}_{22} \end{bmatrix},$$

$$\mathbf{K}_{11} = \begin{bmatrix} \frac{ES}{a} & & & & & \\ & \frac{12EI_z}{a^3} & & & & -\frac{6EI_z}{a^2} \\ & & \frac{12EI_y}{a^3} & & \frac{6EI_y}{a^2} & \\ & & & \frac{GJ_x}{a} & & \\ & & \frac{6EI_y}{a^2} & & \frac{4EI_y}{a} & \\ & -\frac{6EI_z}{a^2} & & & & \frac{4EI_z}{a} \end{bmatrix}$$

$$\mathbf{K}_{12} = \mathbf{K}_{21}^T = \begin{bmatrix} -\frac{ES}{a} & & & & & \\ & -\frac{12EI_z}{a^3} & & & & -\frac{6EI_z}{a^2} \\ & & -\frac{12EI_y}{a^3} & & \frac{6EI_y}{a^2} & \\ & & & -\frac{GJ_x}{a} & & \\ & & -\frac{6EI_y}{a^2} & & \frac{2EI_y}{a} & \\ & \frac{6EI_y}{a^2} & & & & \frac{2EI_z}{a} \end{bmatrix}$$

$$\mathbf{K}_{22} = \begin{bmatrix} \frac{ES}{a} & & & & & \\ & \frac{12EI_z}{a^3} & & & & \frac{6EI_z}{a^2} \\ & & \frac{12EI_y}{a^3} & & -\frac{6EI_y}{a^2} & \\ & & & \frac{GJ_x}{a} & & \\ & & -\frac{6EI_y}{a^2} & & \frac{4EI_y}{a} & \\ & \frac{6EI_z}{a^2} & & & & \frac{4EI_z}{a} \end{bmatrix}$$

Appendix C
Stiffness Matrix Formation Algorithm for a Beam System in Analytical Form

Using the results of Appendix A, an analytical algorithm for the formation of the stiffness matrix for an end-element beam system can be proposed in a general form.

1. In each node i of the system (Figs. A.1 and A.2), it is necessary to define a (general) coordinate system $(x_{io}, y_{io}, z_{io},)$ linked to this node. The end element t of the beam enters into node i with its first or second end. To be more specific, let us assume that this is end 1.
2. Let us determine the rotation matrix $\boldsymbol{\theta}_{ti}$. It depends on the rotation angles $\theta_x, \theta_y, \theta_z$ of the local coordinate system of the t-th beam element and the general coordinate system (x_{io}, y_{io}, z_{io})

$$\boldsymbol{\theta}_{ti} = \boldsymbol{\theta}_1 \boldsymbol{\theta}_2 \boldsymbol{\theta}_3 ,$$

$$\boldsymbol{\theta}_1 = \begin{bmatrix} 1 & 0 & 0 \\ 0 & \cos\theta_x & -\sin\theta_x \\ 0 & \sin\theta_x & \cos\theta_x \end{bmatrix}, \quad \boldsymbol{\theta}_2 = \begin{bmatrix} \cos\theta_y & 0 & \sin\theta_y \\ 0 & 1 & 0 \\ -\sin\theta_y & 0 & \cos\theta_y \end{bmatrix},$$

$$\boldsymbol{\theta}_3 = \begin{bmatrix} \cos\theta_z & -\sin\theta_z & 0 \\ \sin\theta_z & \cos\theta_z & 0 \\ 0 & 0 & 1 \end{bmatrix}.$$

3. Let us calculate the shift matrix $\boldsymbol{\rho}$ for both ends of a beam (if needed):

$$\boldsymbol{\rho}_1 = \begin{bmatrix} 0 & \rho_z & -\rho_y \\ -\rho_z & 0 & \rho_x \\ \rho_y & -\rho_x & 0 \end{bmatrix}_1$$

where ρ_x, ρ_x, ρ_z are the coordinates of the beam ends in the coordinate system (x_{io}, y_{io}, z_{io}). Let us form the transformation matrix for the first end of the beam included in node i

$$\boldsymbol{\Gamma}_{1i} = \begin{bmatrix} \boldsymbol{\theta} & \boldsymbol{\theta}\mathbf{V}_1 \\ \mathbf{0} & \boldsymbol{\theta} \end{bmatrix}_i,$$

Similarly, let us do the same for the transformation matrix $\mathbf{\Gamma}_{2j}$ for the second end of this beam, which is now included in another node j.

5. Let us find the transformation matrix for beam t

$$\begin{bmatrix} \mathbf{K}'_{11} & \mathbf{K}'_{12} \\ \mathbf{K}'_{21} & \mathbf{K}'_{22} \end{bmatrix}_t = \begin{bmatrix} \mathbf{\Gamma}_{1i} & \\ & \mathbf{\Gamma}_{2j} \end{bmatrix}^{-1} \begin{bmatrix} \mathbf{K}_{11} & \mathbf{K}_{12} \\ \mathbf{K}_{21} & \mathbf{K}_{22} \end{bmatrix}_t \begin{bmatrix} \mathbf{\Gamma}_{1i} & \\ & \mathbf{\Gamma}_{2j} \end{bmatrix}.$$

If node i contains several beams, it is necessary to sum up their transformation matrices $\mathbf{K}'_{11}$. Matrices $\mathbf{K}'_{12}$ and $\mathbf{K}'_{21}$ reflect the link between nodes i and j. Matrix $\mathbf{K}'_{22}$ is now related to node j.

The inertia matrix $\mathbf{M}$ is formed similarly. However, very often it is presumed to be a diagonal one.

Appendix D
Stiffness Matrices for a Planetary Reduction Gear Subsystems

The stiffness matrices for a differential gear and epicycle are

$$\mathbf{K}_S = \begin{bmatrix} a_1 & & \\ & a_1 & \\ & & a_2 \end{bmatrix}, \quad \mathbf{K}_{Ep} = \begin{bmatrix} b_1 & & \\ & b_1 & \\ & & b_2 \end{bmatrix}.$$

The linkage matrices for differential gear-satellites are

$$\mathbf{K}_{SC_i} = \begin{bmatrix} -\alpha_i h_1 \cos\gamma & \alpha_i h_1 \sin\gamma & -r_3 \alpha_i h_1 \\ -\beta_i h_1 \cos\gamma & \beta_i h_1 \sin\gamma & -r_3 \beta_i h_1 \\ r_1 h_1 \cos\gamma & -r_1 h_1 \sin\gamma & -r_1 r_3 h_1 \end{bmatrix}.$$

Epicycle-satellite linkage matrix

$$\mathbf{K}_{EpC_i} = \begin{bmatrix} -\alpha_i h_2 \cos\gamma & -\alpha_i h_2 \sin\gamma & r_3 \alpha_i h_2 \\ -\beta_i h_2 \cos\gamma & -\beta_i h_2 \sin\gamma & -r_3 \beta_i h_2 \\ r_2 h_2 \cos\gamma & r_2 h_2 \sin\gamma & -r_3 r_2 h_2 \end{bmatrix},$$

$$\alpha_i = \cos\frac{2\pi(i-1)}{n}, \ \beta_i = \sin\frac{2\pi(i-1)}{n} \ (i = 1\ldots 3)$$

$$\alpha_1 = h_1 + h_7\frac{2}{3}, \ \alpha_2 = h_1 r_1^2 + h_{12}\frac{1}{3}$$

$$b_1 = h_2 + h_9\frac{4}{3}, \ b_2 = h_2 r_2^2 + h_9\frac{4}{3}r_2^2.$$

Satellites stiffness matrices

$$\begin{matrix}\mathbf{K}_{C_1} \\ \mathbf{K}_{C_2} \\ \mathbf{K}_{C_3}\end{matrix} = \begin{bmatrix} (h_1 + h_2)Cos^2\gamma + h_6 & (h_1 - h_2)\cos\gamma\sin\gamma & (h_1 - h_2)r_3 Cos\gamma \\ & (h_1 + h_2)\sin^2\gamma + h_6 & (h_1 + h_2)r_3\sin\gamma \\ symm. & & (h_1 + h_2)r_3^2 \end{bmatrix}.$$

References

1. Akhmetchanov, R.S., Banakh, L.Ya.: Analysis of dynamic properties of regular plate systems. Mashinovedenie **1**, 67–74, Moscow (Russian) (1988)
2. Akhmetchanov, R., Banakh, L.: The vibration interactions analysis in rotor systems of the gas turbine design. Problemy mashinostrojenia i nadezhnost mashin **4**, 29–38, Moscow (Russian) (1992)
3. Aleksandrov, A.V., Lashchennikov, B.J., et al.: Civil Engineering Mechanics. Thin-walled Spatial Systems. Strojizdat, Moscow (Russian) (1983)
4. Ajrapetov, E.L., Genkin, M.D.: Dynamics of Planetary Mechanisms. Nauka, Moscow (Russian) (1980)
5. Argiris, J.: Modern Achievements in Calculation Methods of Constructions with Application of Matrixes. Strojizdat, Moscow (Russian) (1968)
6. Artobolevskij, I.I., Boborovnitskij, J.I., Genkin M.D.: Introduction in Acoustic Dynamics of Machines. Nauka, Moscow (Russian) (1979)
7. Augustaitis, V, Mozura, G. et al.: Automatic Calculation of Machines Oscillations. Mashinostroenie, Leningrad (Russian) (1988)
8. Bakhvalov, N.S.: Numerical Methods. Nauka, Moscow (Russian) (1975)
9. Banakh, L.: Decomposition methods in oscillations of multi- dimensional systems. *Doklady AN, Technical physics* **337**(2), 189–193, Moscow (Russian) (1994)
10. Banakh L.: The weak interactions by mechanical systems oscillations. *Doklady AN., Mechanics*, **337**, 336–338, Moscow (Russian) (1994)
11. Banakh, L.: Decomposition methods for vibration analysis of large mechanical systems. In: Proceedings of the 2nd European Nonlinear Oscillations Conference, Academy of Sci, of Czech Republic Prague, vol. 1, 73–76 (1996)
12. Banakh, L.Ya., Gadjieva, E.G.: Dynamic research of regular and quasyregular systems with rotational symmetry. *Mashinovedenie* **3**, 9–16, Moscow (Russian) (1984)
13. Banakh, L.Ya, Grinkevitch, V.K, Fedoseev, Yu.N.: Analis of mechanisms dynamics with help of symmetry groups. *Mashinovedenie* **2**, 67–71, Moscow (Russian) (1986)
14. Banakh, L.Ya, Fedoseev, Yu.N.: Revealing of independent oscillations in reducer gear owing to its symmetry. In: 12-th Congress in Mech. and mash, Edited by Jean-Pierre Merlet and Marc Dahan. Besancon, France. (2007)
15. Banakh, L.Ya., Perminov, M.D.,Sinev, A.V. et al. (1973): Methods of calculation of stiffness, inertia and damping matrixes for complicated spatial systems. In: Vibroisolation machines and vibro- defense of man – operator. Moscow, Nauka, 67–81(Russian)
16. Baranov A.V.: Elements of the theory of systems with the distributed connections. Bulletin of the Moscow State University, series 3, N 5(1965)
17. Bathe, K., Wilson, E.L.: Numerical Methods in Finite Element Analysis. Prentice-Hall, Inc,. (1976)
18. R. Bellman. Introduction to Matrix Analysis. MCGraw Hill Company, Inc., New York, (1960)

19. Benfild, J., Ruda, C.: Oscillations research of constructions based on oscillations forms of separate subsystems. *AJJJ* **9**(7), 52–61 (1971)
20. Biderman, V.L.: Theory of Mechanical Oscillations. Vysshaja shkola, Moscow (Russian) (1980)
21. Bisplinghoff, R.L., Ashley, H. et al.: Aeroelasticity. Addison-Wesley Publishing Company, Inc., Cambridge (1955)
22. Bleich, F.: The Equations in Finite Differences for Constructions Statics, Gostechizdat, Kharkov (Russian) (1933)
23. Bolotin, V.V., Novichkov Yu. N.: The Mechanics of Multilayered Designs. Mashinostroenie, Moscow (Russian) (1980)
24. Bolotin, V.V.: Theory of natural frequencies distribution for elastic bodies and its application to problems of accidental oscillations. Prikladnaya mechanika, **8**(4), 3–29 (Russian) (1972)
25. Brillyuan, L., Parodi, M.: Distribution of Waves in Periodic Structures. Inostrannaya literature, Moscow (Russian) (1959)
26. Buryshkin, M.L., Gordeev, V.I.: Effective Methods and Calculation Programs for Symmetric Constructions. Budivelnik, Kiev (Russian) (1984)
27. Collatz L.: Eigenwertaufgoben mit technishen Anwendungen. Akademische Verlagsgesselshaft Geest & Portig K.-G., Leipzig (1963)
28. Creig, R.R., Bampton, M.K.: Coupling of substructures at dynamic analysis. *AJJJ* **6**, 113–121 (1968)
29. Dinkevich, S.Z.: Calculation of Cyclic Constructions. A Spectral Method. Strojizdat, Moscow (Russian) (1977)
30. Dimentberg, F.M.: Flexural Vibrations of Rotating Shafts, Butterworth, London, United Kingdom (1983)
31. Dondoshanskii, V.K.: Calculation of Oscillations of Elastic Systems on Electronic Computers. Mashinostroenie, Moscow-Leningrad (Russian) (1965)
32. Feder J.: Fractals. Plenum Press, NY (1991)
33. Fedoseev, Yu., Pusakina A.: Analysis of dynamics of mechanical systems with help of symmetry groups. In: Problemy masinostrojenia i nadezhnost mashin, 1, Moscow (Russian) 26–34 (2009)
34. Filin, A.P.: Matrixes in Static of Rod Systems, Moscow (Russian) (1966)
35. Filin, A.P., Tananajko O.D. et al.: Algorithms of Resolving Equations of Rod Systems. Strojizdat, Moscow (Russian) (1983)
36. Fillippov, A.G.: Oscillations of Mechanical Systems. Naukova Dumka, Kiev (Russian) (1965)
37. Fomin,V.M.: Application of the theory groups representation to definition of frequencies and forms of oscillations for rod systems with the given symmetry group. In: Distributed Control of Processes in Medium, vol 1, pp. 144–151, Naukova Dumka, Kiev (Russian) (1969)
38. Fulton, W., Harris, J.: Representation Theory. A first Course. Springer Verlag (1991)
39. Gajan, R.: Reduction of stiffness and mass matrixes. Rocket Engineering and Astronautics **2**, 287–299 (1965)
40. Gallagher R.H.: Finite Elements Analysis. Fundamentals. Prentice-Hall, New Jersey (1975)
41. Gantmacher, F.R.: Theory of Matrixes. Nauka, Moscow (Russian) (1967)
42. Gelfond, A.O.: Calculation of finite differences. Nauka, Moscow (Russian) (1967)
43. Grinkevitch, V.K.: Vibrations of elements of planetary reducer. In: Static and Dynamic of Mechanisms with Tooth, Nauka, Moscow, 102–113 (1974)
44. Grodko, L.N.: About Small Oscillations of Mechanical Systems having Circular Symmetry. Izv. AN SSSR, 1, Moscow (Russian) (1967)
45. Ivanov, V.P.: About some vibrating properties of the elastic systems, having cyclic symmetry. In: Durability and Dynamics of Aviation Engines, vol. 6, Mashinostrojenie, Moscow (Russian) pp. 113–132 (1971)
46. Ivanov, V.P.: Oscillations of Impellers of Turbomachines. Mashinostrojenie, Moscow (Russian) (1984)

47. Ivovitch, V.A.: Transition Matrixes in Dynamics of Elastic Systems. Mashinostrojenie, Moscow (Russian) (1969)
48. Kandidov, V.P., Chesnokov, S.S.: Application of Finite Elements Method for Calculation of Transversal Oscillations for One-dimensional Systems. Vestnik of Moscow State University, series 3, 4, (Russian) (1971)
49. Kartvelishvili, N.A., Galaktionov, J.I.: Idealization of Complicated Dynamic Systems. Nauka, Moscow (Russian) (1976)
50. Kempner, M.L.: Statistical analysis of shaft line torque of internal, combustion engines in resonanse conditions. In: Engine construction, **9** (1989)
51. Kempner, M.L.: To studies of random oscillations of aircraft engine blades. In: Problems of strength, **1** (1984), 65–67
52. Kempner, M.L.: Application of dynamic compliances method for calculation of cranked shafts oscillations. In: Dynamics and Durability of Cranked Shaft, vol. 2. Ed. AN SSSR, Moscow 34–48 (Russian) (1950)
53. Kempner, M.L.: Dynamic compliances and stiffness methods for calculation of bending oscillations of elastic systems with many degrees of freedom. In: Transversal Oscillations and Critical Speeds, Ed. AN SSSR, Moscow (Russian) pp. 78–130 (1951)
54. Kempner, M.L.: Application of topology methods for calculation of the connected oscillations of gasturbine engines. In: Problems of Dynamic and Durability of Engines. Leningrad (Russian) (1973)
55. Kempner, M.L., Tselnits, D.S.: Oscillations of Systems with Cyclic Symmetry. In: Trudy MIIT, vol. 476, Moscow (Russian) 25–32 (1975)
56. Kempner, M.L., Nesterova S.V.: Research of Connectionless for Torsion and Longitudinal Oscillations of Cranked Shaft. Mashinivedenie, 1, Moscow (Russian) (1975)
57. Kempner, M.L.: Dynamic compliances of quasiregilar one-dimensional system. In: Trudy MIIT, vol. 466, Moscow (Russian) 212–216 (1974)
58. Kempner, M.L.: Transfer Matrixes of quasiregilar systems. In: Trudy MIIT, vol. 466, Moscow (Russian) pp. 209–212 (1974)
59. Kempner, M.L., Soboleva O.N.: Oscillations of Cylindrical Shell Supported with Longitudinal Ribs with Asymmetrical Structure. In: Novosti Vysshei shkoly, Mechanical engineering, 7, Moscow (Russian) pp. 29–32 (1975)
60. Kempner, M.L., Efremova, B.T.: Oscillations of Shrouded Blades of Turbomashines. In: Trudy MIIT, vol. 509, Moscow (Russian), pp. 42–52 (1977)
61. Khammermesh, M.: The Theory of Groups and its Application to Physical Problems. Mir, Moscow (Russian) (1966)
62. Khejl, A.L., Mierovitch, L.: A method of improvement of representation of discrete subconstruktions in dynamic synthesis of constructions. Aerokosmicheskaja technika, **1**(3), 124–135 (Russian) (1983)
63. Koenig, G., Blackwell,: Theory of Electromechanical Systems. Izd. Energy, Moscow (Russian) (1965)
64. Kollatz, L.: Eeigenvalues Problems. Nauka, Moscow (Russian) (1966)
65. Kostyuk, A.G.: Dynamics and Strength of Turbo Machines. Mashinostrojenie, Moscow (Russian) (2000)
66. Koutsovasilis, P, Beitelschmidt, M.: Model Reduction of Large Elastic Systems. A Comparision Study on the elastic Piston Rod. In: 12-th Congress in Mech. and mash Sc.Besancon, Edited by Jean-Pierre Merlet and Marc Dahan France (2007)
67. Kron, G. Kron, G.: Research of complicated systems in parts. Diakoptika. Nauka, Moscow (Russian) (1972)
68. Kushner, F.: Rotating Component Modal Analysis and Resonance Avoidance Recommendations. In: Proceedings of the thirty-third Turbomachinery symposium, Turbomachinery Laboratory. Texas A&M University. College Station. Texas, 143–161 (2004)
69. Landa, P.S.: Nonlinear Oscillations and Waves in Dynamical Systems. Dordrecht-Boston-London: Kluwer Academic Publ. (1996)

70. Lashchennikov, B.J., Dolotkazin, D.V.: About application of a finite elements method in problems of waves distribution. The Building Mechanics and Calculation of Constructions **6**, 52–54 (Russian) (1983)
71. Lyubarskii G.Ya.: Theory of Groups and its Application in Physics. Gostehizdat, Moscow (Russian) (1957)
72. Mac-Neal, R.H.: Hybrid Method of Component Mode Synthesis. Computer and Structures, **1**, 581–601 (1971)
73. Mac-Donald, D.: Problem of natural oscillations for the supported cylindrical shells. Rocket Engineering and Astronautics, 8, Mir, Moscow (Russian) (1970)
74. Mandelbrot, B.B.: Fractals and Multifractals, Selecta, vol. 1. Springer, N.Y. (1991)
75. Mandelbrot, B.B.: Fractal Geometry of Nature. Institute of Computer Researches, Moscow (Russian) (2002)
76. Mandelstam, L.: The Lectures on the Oscillations Theory. Nauka, Moscow (Russian) (1972)
77. Manevitch, L.I., Andrianov, I., et al.: Mechanics of Periodically Heterogeneous Structures. Berlin: Springer-Verlag (2002)
78. Maslov, V.P., Rimskii-Korsakov, A.V.: Plane wave in a plate with parallel ribs. In: Vibration and Noise, Nauka, Moscow 29–38 (Russian) (1969)
79. Miller, Ch., Sce, A.A.: Dynamic Reduction of Structural Modes. J. Struct. Div., **106**(10), 2097–2108 (1980)
80. Nagaev R.F., Chodzhaev K.Sh.: Oscillations of Mechanical Systems with Periodic Structure FAN Uzbek SSR, Tashkent (Russian) (1973)
81. Natanson, V.Ya.: Oscillations of Shaft. Oborongis (Russian) (1954)
82. Nayfe, A.A: Perturbation Methods. Mir, Moscow Moscow (Russian) (1976)
83. Novozhilov, V.V.: Theory of Thin Shells. Sudpromgis, Leningrad (Russian) (1962)
84. Ovsyannikov, L.B.: Group Analysis of Differential Equations. Nauka, Moscow (Russian) (1978)
85. Osipov, V.P., Stankevich, A.I.: Calculate - experimental method of vibration forecasting for complicate mechanical systems. In: Dynamics and Identification of Mechanical Systems, pp. 19–27, Ivanovsky Energ. Inst. Ivanovo (Russian) (1985)
86. Palmov, V.A., Pervozvanskii, A.A.: About computing features of matrix calculation methods for oscillations. Trudy LPI, 210, Dynamic and Durability of Machines, Mashgis, Moscow-Leningrad (Russian) (1960)
87. Panovko, Ya.G.: Foundation of the Applied Theory of Elastic Oscillations. Mashgis, Moscow (Russian) (1967)
88. Panovko, Ya.G., Gubanova, I.I.: Stability of Elastic Systems Oscillations. Nauka, Moscow (Russian) (1979)
89. Pain, H.J.: Physics of Vibrations and Waves. John Willey. London, (1976)
90. Paintgen, Kh.O., Rikhter, P.Kh.: Beauty of Fractals, Mir, Moscow (Russian) (1993)
91. Painter, P., Coleman, M, et al.: The theory of Vibrational Spectroscopy. A Wiley-Interscience Publication, John Wiley, New York (1986)
92. Perminov, M.D., Petrov, V.D.: Research of oscillations of complicate systems by a partitions method. In: Dynamics and Durability of Elastic and Hydro-elastic Systems, Nauka, Moscow (Russian), 41–52 (1975)
93. Pervozvanskii, A.A.., Gaizgory, V.: Decomposition, Aggregation and the Approximate Optimization, Nauka, Moscow (Russian) (1979)
94. Postnov, V.A., Kharkhurim, I.A.: The Finite Elements Method in Calculations of Ship Constctions. Sudostroenie, Leningrad (Russian) (1974)
95. Postnov, V.A.: The Method of Superelements in Calculations of Engineering Constructions. Sudostroenie, Leningrad (Russian) (1976)
96. Przemieniecki, I.S.: Theory of Matrix Structural Analysis. N.Y.: Mc. Graw Hill Book Company (1968)
97. Rabinovich, M.I., Trubetskov, D.I.: Introduction in the Theory of Oscillations and Waves. Nauka, Moscow (Russian) (1984)

98. Rivin, E.I.: Dynamics of Machines Drives. Mashinostrojenie, Moscow (Russian) (1966)
99. Rubin, S.: Refinement of oscillations forms for dynamic calculations of constructions. *JAAA*, **13**(8), 35–49 (1975)
100. Segal, A.I.: Rotating symmetry at rotating loadings. In: Researches Under the Theory of Constructions. Gostechizdat, vol. 8, pp. 27–34 Moscow (Russian) (1957)
101. Seshu, S., Rid, M.B.: The Graphs and Electric Circuits. Vysshaja shkola, Moscow (Russian) (1971)
102. Smolnikov, B.A.: Calculation of free oscillations for closed frame systems with cyclic symmetry. In: Dynamics and Durability. Trudy LPI, 210, Mashgis (Russian) (1960)
103. Steindel, A., Troger, H.: Dimension Reduction: A Key Concept in Dynamics. ENOC-2008,Saint Peterburg, Russia (2008)
104. Strelkov, S.P., Strelkov, S.P.: Introduction in the Theory of Oscillations. Nauka, Moscow (Russian) (1964)
105. Strutt, J.W (Lord Rayleigh): The Theory of Sound. MacMillan Co. Limited, London, vol. 1, 2, (1926)
106. Timoshenko, S.P.: Oscillations in Engineering. Nauka, Moscow (Russian) (1967)
107. Trojevich, V.M., Kunar, R., et al.: The reduction of degrees of freedoms in dynamic analysis on complicate three-dimensional structures.. In: Armer, G.S.T., Garas, T.V.(eds) The Use of Modal Analysis, Construction Press, pp. 178–181 (1982)
108. Vejts, V.L., Kolovskij, M.Z., et al.: Dynamics of Controlled Machine Aggregates. Mashinostrojenie, Moscow (Russian) (1984)
109. Vibrations in Technique. Hand Book: Mashinostrojenie, Moscow (Russian) v.1 Ed. Bolotin V.V., v.3 Ed. Frolov K.V. (1980)
110. Vulfson J.I.: Vibroactivity of Branched and Ring Structured Mechanical Drives. Hemisphere Publ. Corpor.: New York, Washington, London (1998)
111. Wilkinson, J.: An Algebraic Problem of Eigenvalues. Nauka, Moscow (Russian) (1970)
112. Zhermen, P.: Mechanics of Medium. Mir, Moscow (Russian) (1965)
113. Zhuravlev, V. F., Klimov, D.M.: Applied Methods in the Theory of Oscillations. Nauka, , Moscow (Russian) (1988)
114. Zienkievicz, O.: Finite Elements Method in Engineering. Mir, Moscow (Russian) (1975)
115. Zlokovich, J.: Theory of Groups and G-vector Spaces in Oscillations, Stability and a Static of Constructions. Strojizdat, Moscow (Russian) (1977)

Index

Note: The letters 'f' and 't' following locators refer to figures and tables respectively.

The manufacturer's authorised representative in the EU is Springer Nature Customer Service Centre GmbH, Europaplatz 3, 69115 Heidelberg, Germany. If you have any concerns regarding our products, please contact ProductSafety@springernature.com

Printed and bound by CPI Group (UK) Ltd, Croydon, CR0 4YY
01/10/2026
02225968-0006